Tropical Plants
for Home and Garden

Tropical Plants
for Home and Garden

William Warren

Photographs by Luca Invernizzi Tettoni

with 449 color illustrations

Thames & Hudson

A Note on Plant Names

The names of tropical plants present certain problems to the non-professional, not least because of the changes in many that have taken place over the years as botanists refine their classifications. Some of these are controversial and confusing to a layman. *Cordyline*, for example, which thrives in wet, tropical regions was once included in the *Liliaceae*, or lily, family; it has now been moved to the family of *Agaveaceae*, which includes the Century Plant and is found in warm, dry regions. The climbing Aroid now known as *Epipremnum pinnatum* 'Aureum' (popularly as Golden Pothos) has, at various times, been listed under the genera *Pothos*, *Scindapsus*, and *Rhaphidophora*. In various reference works the Torch Ginger can be found listed as *Phaeomeria magnifica*, *P. speciosa*, *Nicolaia elatior*, and, apparently the current choice, *Etlingera elatior*; in two books, both published in 1995, one offers the first of these names and the other gives the last.

In this book, every effort has been made to conform with the latest designation. Errors, however, have almost certainly crept in, will undoubtedly be reported by expert readers, and will be corrected in any future edition.

First published in Great Britain as *Tropical Garden Plants*

© 1997 Thames & Hudson Ltd, London

First published in hardcover in the United States of America in 1997 by Thames & Hudson Inc., 500 Fifth Avenue, New York, New York 10110

thamesandhudsonusa.com

First paperback edition 2006

Library of Congress Catalog Card Number 97-60248

ISBN-13:   978-0-500-28341-7
ISBN-10:   0-500-28341-9

Printed in China by Toppan Printing Co.

Contents

Introduction

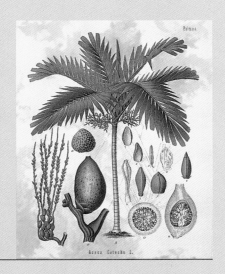

Almost everyone is familiar with the story of Captain Bligh, HMS *Bounty*, and the mutiny led by Fletcher Christian. Less well known, perhaps, is exactly what brought Bligh to the island of Tahiti towards the end of 1788, why he and his men spent such a long time in those idyllic tropical surroundings, and how such factors made the famous mutiny almost inevitable.

Bligh was sent to the South Seas on the recommendation of Joseph Banks, the wealthy and enthusiastic amateur botanist who had accompanied Captain James Cook on his first pioneering voyage nearly twenty years earlier. His mission was to collect 1,000 seedlings of *Artocarpus altilis*, a native tree popularly called the Breadfruit, and transport them halfway around the world to British colonies in the West Indies. The tree produced huge, nutritious fruits as often as three times a year, and it was thought they might provide an inexpensive food supply for African slaves then working on Caribbean plantations.

It took time, though, to collect and germinate the seeds, and almost five months for the seedlings to reach a sufficient maturity to survive the long voyage ahead. "No large group of Europeans", Alan Moorehead has written, "had remained so long on the island before, and the attachments formed by the Bounty's crew with the Tahitian women were something more than those of a sailor's spree. Every man had his girl, and when they came to sail away many of them found the loss of their companions unendurable. That is why Fletcher Christian and his followers so treacherously rebelled when the Bounty reached the Friendly Islands: they wanted to return to Tahiti and the easy life."

Bligh eventually reached England, crossing the Pacific in an open boat. He returned to Tahiti in 1792, collected his Breadfruit seedlings, and carried them safely to the West Indies, where the slaves did not find them nearly so appealing as the Polynesians had.

Left   **A plantation of coconut palms in Southern Thailand**
Above **Botanical drawing of *Areca catechu***
The Betel Nut Palm

BOTANICAL GARDENS. SINGAPORE.

Extension of original cutting on an old Para rubber

Singapore. Bot. Garden. Victoria Regia.

1 **Specimen palm trees at the Singapore Botanic Gardens**

2 *Victoria regia* (now *Victoria amazonica*)
The Giant Water Lily

3 *Hevea brasiliensis*
The Para Rubber Tree

As this dramatic incident reveals, the movement of tropical plants around the world was already a significant feature of European exploration and had been from the earliest days. On his second voyage to the New World, in 1493, Christopher Columbus was given an odd, cone-shaped fruit by the natives of Guadaloupe. It reminded the Spanish visitors of a pine cone, so they gave it the name of "pina", which later became pineapple and, to botanists, *Ananas comosus*.

Originally a native of Brazil, this member of the *Bromeliaceae* family had been cultivated throughout the New World tropics long before the arrival of Europeans. Finding it "delicious and freshing to the taste", Columbus brought some samples back to the King, and later Spanish and Portuguese explorers began its widespread distribution. By 1548 it was under cultivation in Madagascar, by 1590 in India; following its introduction to Hawaii, vast pineapple plantations became a major industry that dominated island economy until recent years when, due to labour costs, they moved to such places as the Philippines, Thailand, and Malaysia.

As these examples show, the initial focus was on plants that might have some commercial (often culinary) possibilities both in Europe and in tropical colonies then being established in remote places like India and the Spice Islands of Indonesia. Another of Columbus's discoveries was *Capsicum*, the chilli pepper, an established part of the diet in South America and Mexico for at least 7,000 years. He called it pimento, an adaptation of the Spanish word for black pepper, one of the objects of his explorations; but the first seeds sent back were planted not for seasoning food but for the ornamental appeal of their bright red and orange fruits. *Capsicum* probably reached India with the Portuguese (they were recorded in Goa in the middle of the 16th century, but there is no mention in ancient Sanskrit books on food) and subsequently became such an essential part of Asian cuisine that their foreign origin was soon forgotten. The British, for example, got *Capsicum* first from India and then reintroduced the spice to their colonies in the New World.

Other famous economic plants that found new homes and created new industries were the Oil Palm (*Elaeis guineensis*), originally from Africa, and the Para rubber tree (*Hevea brasiliensis*), which transformed Malaya when it was introduced from its native Brazil, first through Kew and then the Singapore Botanic Gardens.

But more than 200,000 species flourish abundantly in the fertile tropics, and many were also carried far from their original habitat, not because of their commercial possibilities but simply because they were extraordinarily beautiful or because they had rare curiosity value, and sometimes for more complex reasons.

A millennium ago, when Polynesians came to Hawaii in giant outrigger canoes, their cargo included the green *Cordyline*, the "ti" plant, regarded as a symbol of divine power and woven into leis for the highest-ranking priests; at some later time, the red *Cordyline* was introduced to the islands in the same way. Exactly when the wonderfully fragrant *Plumeria*, or frangipani, began its wanderings is uncertain, but probably either the Portuguese or the Spanish – perhaps both – carried it from its New World origins to distant Asia, where it soon became closely identified with various religions and acquired an extensive collection of new local names. It was in

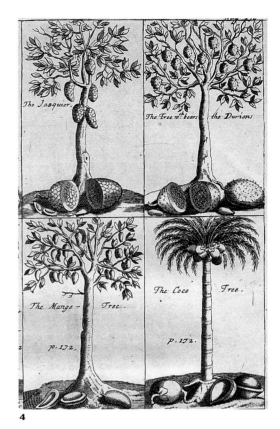

4

4 **Plants with culinary possibilities: "The Jacquier; The Mango Tree; The Tree which bears the Durions; The Coco Tree."**

1

2

Brazil that the French explorer Louis Antoine de Bougainville came across the dazzling creeper, magenta-coloured at the time, that was later named after him and that is now almost universally seen in both tropical and sub-tropical gardens.

Particularly in the 19th century, there was keen competition to acquire new and ever odder specimens to display in the glasshouses of Great Britain and Europe. Nurseries like Veitch in Exeter (later London) sent out their own plant hunters, as did great botanic gardens and wealthy private collectors. A number of people can claim credit for discovery of *Victoria amazonica*, the gigantic South American water lily named after the British Queen, but it was first induced to flower in captivity at Chatsworth, home of the Duke of Devonshire. Strange new orchids were brought from the jungles of Malaya, Borneo, Java, and the Philippines, along with other desirable ornamentals such as *Dracaena*, *Ixora* and *Codiaeum* or *Croton*, with exotically patterned leaves. F.W. Burbidge, a Veitch plant hunter, went to wildest Borneo for the sole purpose of collecting *Nepenthes*, the weird Pitcher Plant, which traps insects and then slowly digests them in a sinister cup-like appendage. He came back with 47 species, all new to the non-tropical world, which soon spread to glasshouses in Europe and America.

Once removed from their native habitat, many of these specimens underwent considerable changes. The most popular ones – *Codiaeum*, for instance, as well as *Cordyline*, *Hibiscus*, *Plumeria*, and members of the Aroid family – were hybridized into an infinite variety of cultivars that differed greatly from the original in the shape and colour of both flowers and leaves. Some of this work was done actually in the tropics, at famous research centres like Bogor in Java, but much of it took place far away in British, French, Belgian, or American nurseries, mainly to keep up with the changing tastes of house plant fanciers.

Aldous Huxley, on a visit to Southeast Asia in the 1920s, was responding to this enthusiasm for the strange and beautiful when he observed that "the special and peculiar charm of tropical botany is that you can never be quite sure that it isn't zoology, or arts and crafts, or primitive religion. There are lilies in Malaya whose petals have become attenuated to writhing tentacles, so that they dangle on their stalks like perfumed spiders. There are palms whose fruits are vegetable porcupines…. There are orchids in Singapore that might be pigeons, and others from which one recoils instinctively as though from the head of a snake. The gardens of the equator are full of shrubs that bloom with votive offerings to the Great Mother, and are fruited with coloured Easter eggs, lingams and swastikas. There are trees whose stems are fantastically buttressed to look like specimens of a late and decadent Gothic architecture…. There are red varnished leaves and leaves of shiny purple that look as though they were made of American cloth or patent leather. There are leaves cut out of pink blotting-paper; leaves whose green is piped with lines of white or rose in a manner so sketchily elegant, so daring, so characteristically 'modern', that they are manifestly samples of the very latest furniture fabrics from Paris."*

* Jesting Pilate (Chatto and Windus Ltd., 1926)

3

1 ***Nepenthes*** The Pitcher Plant
2 **_Euphorbia pulcherrima_** The Poinsettia **from Mexico. Now a common house plant.**
3 **A sinister orchid from Borneo**
4 **Epiphytic tropical orchids shown in their natural habitat**

4

Both dispersal and hybridization continue today, at an even more rapid rate thanks to improvements in transportation and laboratory skills. To cite merely one random example, less than a decade ago gardeners in Thailand had perhaps half a dozen varieties of *Heliconia* to choose from. In that brief space of time this flamboyant relative of the banana with its brightly coloured flower bracts has enjoyed a boom and countless new species have appeared on the market, some only recently collected from the South American jungles where they grow wild. A similar explosion of interest is also taking place elsewhere: the first comprehensive identification guide to *Heliconia*, published only in 1991, is already regarded as out of date and another one is being prepared to include additional discoveries and hybrids.

The results are apparent to anyone who makes a garden tour of tropical countries. Whether in Bangkok or Barbados, Honolulu or Miami, non-native species have become established features of the landscape, often in the distant past. To those already mentioned, we can add such now-standard specimens as the Flame Tree, or Flamboyant (*Delonix regia*), and the so-called Traveller's Palm (*Ravenala*), both originally from the island of Madagascar, as well as the Bird of Paradise (South Africa), the Red Ginger (Malaysia), *Allamanda* (South America), the Bottlebrush (Australia), the Golden Shower Tree (South Asia), and the red-stemmed Sealing-wax Palm (Sumatra).

Nor can such plants be seen and enjoyed only in what is conventionally regarded as a tropical climate, that is, a place of uniformly high temperatures and humidity, perpetually refreshed by rain. Many of them attracted their first popularity as much-admired specimens in glasshouses, living rooms, or the lobbies of public buildings, and they continue to be seen in such settings; far more people, it is safe to say, are familiar with *Codiaeum, Dieffenbachia*, and the Pygmy Date Palm as decorative, lovingly tended potted specimens than have ever seen them as integral parts of a garden design.

Moreover, many landscape designers are discovering that some species are remarkably hardy even outdoors in non-tropical settings. One who lives and practises near San Francisco, where temperatures can drop to freezing for a short time during the winter months, uses them often in his work. "In Northern California", he notes, "most houses have large windows and I use the radiated heat from these to keep my tropicals going through the cold spells." In such comparatively protected areas he has found that he can grow *Heliconia, Canna*, numerous *Begonia* hybrids, *Russelia equisetiformis* (the lovely Coral Plant), tree ferns, and many Ginger species. Other gardeners elsewhere are discovering that many fast-growing tropicals can be planted outside when the danger of frost has passed and enjoyed for at least a substantial part of the year.

The world of tropical plants is one of infinite beauty and variety, of colours and forms rarely encountered in temperate gardens. Whether for actual use, or merely for the wonder they inspire, they provide a source of endless horticultural pleasure and interest.

1 Ornamental Trees

Many of the most ornamental tropical shrubs and creepers are familiar to plant lovers in temperate countries thanks to their widespread use in greenhouses and as potted specimens. Less well known, perhaps, because of their size, are the numerous ornamental trees that play such a fundamental role in garden landscapes, bringing beauty in the form of often spectacular flowers and also a wide range of shapes and textures.

Over the past three or four centuries such trees have spread from their original habitats throughout the tropics, so that today's gardener has an extensive variety of choices from places he may never have seen. The famous Flamboyant, or Flame Tree (*Delonix regia*), for example, came from Madagascar, whereas the Rain Tree (*Samanea saman*) and the frangipani (*Plumeria*), both now grown almost everywhere, originated in tropical America. The showy orange-red flowers of the Tulip Tree (*Spathodea campanulata*) were first seen in West Africa, the Bottlebrush (*Callistemon*) in Australia, the Jacaranda (*Jacaranda mimosifolia*) in Brazil, and the dazzling Golden Shower (*Cassia fistula*) in southern Asia. Adapted to local conditions, sometimes hybridized into new varieties, these and countless others have become truly international, as basic to the gardens of Southeast Asia as they are to those of Honolulu and Miami.

Left and above **Cassia siamea**

15

4

Cassia

Botanical Family: *Leguminosae*

Cassia is a large genus with some 500 species, among which are a number of highly attractive flowering trees. To many tropical gardeners, the most beautiful is *C. fistula*, popularly called the Golden Shower Tree or the Indian Laburnum. Native to South Asia, this is a medium to large tree with ovate, pointed leaflets; when these drop, usually in the dry season, masses of bright gold flower clusters appear on almost every branch. This has been crossed with *C. javanica*, which has pink and white flowers, to produce a wide range of "Rainbow Shower" hybrids in blends of pink, apricot, and yellow. *C. multijuga* and *C. siamea* are also yellow-flowering, though less spectacular, while *C. bakeriana*, a native of Thailand, is pink-flowering.

All these species of *Cassia* like full sun and well-drained soil and to bloom profusely require a dry season.

2

3

1 **Cassia fistula** Golden Shower Tree	3 **Cassia bakeriana**
2 **Cassia javanica x C. fistula**	4 **Cassia suratensis**

Lagerstroemia

Pride of India, Crape Myrtle

Botanical Family: *Lythraceae*

Lagerstroemia is a genus with 30 species, several of which are among the most popular flowering trees used both in gardens and along avenues. One of the largest is *L. speciosa*, a native of India, which can grow up to 24 metres and has long, pointed leaves which fall during the dry season; as new foliage appears, so do upright panicles of large, showy flowers that range from pale to deep mauve, with occasional white or pink varieties. *L. indica*, the Crape Myrtle, is a smaller, upright tree, growing to around 7 metres, with similar purplish-pink flowers. Both species are slow-growing but flower profusely even when quite young.

Propagation is by seeds or air-layering. Some spraying is necessary for young trees, which seem to attract a number of leaf-eating insects.

2

1

3

Acacia auriculiformis ▼

Wattle

Botanical Family: *Leguminosae*

This is a very fast-growing, bushy tree which can attain a height of up to 10 metres. On some specimens, the crown is dense and rounded; on others, especially as they age, it is much looser. What appear to be leaves are actually phyllodes, flattened leaf stalks; the fragrant, mimosa-like flowers appear frequently on short spikes and are followed by tightly coiled seed pods. Though this species is useful because of its rapid growth and tolerance of most conditions, it requires a good deal of maintenance due to its almost continuous leaf fall.

Better suited to small gardens are *Acacia cincinnata*, which is lower and bushier and has smaller yellow flowers, and *A. mangium*, a medium-sized tree that produces fluffy white flowers on long spikes. *Acacia* is easily propagated by seeds.

Brownea ▲

Botanical Family: *Leguminosae*

This genus of small to medium-sized spreading trees, native to tropical America, is popular with gardeners because of the showy red to red-orange flowers produced along the branches and trunk, seen to best advantage when standing below the tree. The leaves are long and narrow; when young they are limp and pendulous, turning darker green as they harden. *B. grandiceps*, popularly called the Rose of Venezuela, has exceptionally large red flower clusters, as do *B. capitella* and *B. coccinea*. *B. ariza* grows more vigorously but has smaller flowers.

Brownea likes partial shade and humid conditions. Seeds germinate easily, but rarely appear in some areas where air-layering (or marcotting) is the preferred method of propagation.

1 **Lagerstroemia speciosa**	3 **Brownea grandiceps**
2 **Lagerstroemia indica**	4 **Acacia auriculiformis**

4

Ficus

Botanical Family: *Moraceae*

Ficus is a genus containing more than 100 species. One, *F. carica*, is the familiar fig tree grown for its fruit even in fairly cold climates, while many others are tropical, either wild or used in gardens as ornamentals.

Perhaps most familiar to indoor gardeners in Europe and the United States is *F. elastica*, commonly known as the Rubber Plant. In fact, it was once an important source of rubber in its native South Asia until the late-19th-century introduction of *Hevea brasiliensis*, which later filled vast plantations in Malaysia, Thailand, and Indonesia. With its large, leathery, glossy green leaves, *F. elastica* has long been popular as a house plant, requiring little care and tolerating even air-conditioning. In the tropics, however, it becomes a giant many metres tall, with invasive, strangling aerial roots that wrap around neighbouring trees and walls, it therefore requires frequent pruning to keep it under control. There is also an attractive variegated form.

1 **Ficus benghalensis**
Banyan Tree

2 **A variegated form of**
Ficus elastica Rubber Plant

3 **Ficus benghalensis**
Banyan Tree

1

2

3

4

6

5

4 *Ficus religiosa*   6 *Ficus celebensis*

5 *Ficus benjamina*   7 *Ficus microcarpa*

Another species popular as an indoor plant – and which also attains an impressive size in the tropics – is *F. benjamina*, the Weeping Fig, which has small, shiny leaves and drooping branches. Very decorative in a large garden where its shape can be fully appreciated, it should not be planted too near a house because of its extensive root system, which can block drains and undermine foundations. Like *F. elastica*, it is a strangler, and seeds deposited by birds in a hollow in another tree will grow rapidly, put out roots, and eventually kill the host plant.

F. benghalensis, the Banyan Tree, has perhaps the most spectacular aerial roots, which drop from the branches and form trunk-like columns that often extend far from the parent tree like a dim-lit forest. In India and Southeast Asia, the twisting roots of these and similar *Ficus* trees are popularly supposed to be the abodes of spirits and propitiatory offerings are placed nearby.

F. religiosa, a medium to large species with distinctive heart-shaped leaves, is believed by Buddhists to have been the tree under which the Buddha attained enlightenment and is therefore regarded as sacred. Nearly every Buddhist temple contains a specimen somewhere in the compound; one at Amarapura in Sri Lanka is said to be more than 2,000 years old and a descendant of the one at Buddha Gaya in India where the miracle took place.

Other ornamental Ficus include *F. lyrata*, a small tree from tropical West Africa with lyre-shaped leaves; *F. celebensis*, which has branches that droop elegantly; *F. rubiginosa* 'Variegata', with green-and-white leaves; *F. triangularis*, also from Africa, with small triangular leaves and bright orange fruits; and *F. microcarpa*, a compact tree that is often clipped into formal shapes.

7

2

Tabebuia

Botanical Family: *Bignoniaceae*

Tabebuia is a genus containing around 100 species, all native to tropical America but many widely used in hot-weather gardens everywhere. In regions with a sharp division between wet and dry seasons, *T. rosea*, the Pink Trumpet Tree, annually sheds its foliage and the leafless branches are covered with clusters of pale to deep pink flowers. The smaller *T. pentaphylla*, also pink, blooms less profusely but does so more or less continuously without shedding its leaves.

Two other decorative members of the family are *T. donnell-smithii*, the Gold Tree, which after its annual leaf drop becomes a mass of bright golden-yellow flowers, and *T. caraiba* (formerly *T. argenta*), the Silver Trumpet Tree, which has handsome silver-grey leaves and bright yellow tubular flowers.

1

1 **Tabebuia caraiba**
Silver Trumpet Tree

2 **Tabebuia rosea**
Pink Trumpet Tree

Amherstia nobilis

Botanical Family: *Leguminosae*

Named after the wife of a Governor-General of India, this has been described as the most spectacular of all tropical flowering trees; in 1853 the Duke of Devonshire sent a gardener all the way to the subcontinent to obtain a specimen for his collection. It is native to Burma, a medium to large evergreen tree on which the new leaves are limp and reddish purple, later turning green. Once a year in the dry season it produces clusters of large, coral-coloured flowers that hang down like chandeliers.

Amherstia produces few seeds and is somewhat difficult to grow in gardens, especially at moist lower elevations. It prefers forest conditions and needs a pronounced dry spell to bloom at its best.

3 *Amherstia nobilis*
Pride of Burma

3

1

2

3

Saraca

Botanical Family: *Leguminosae*

Several species of *Saraca* have outstandingly beautiful flowers. One of the best for gardens is *S. thaipingensis*, which grows to several metres in height with large showy masses of orange flowers on the old wood; new leaves appear as limp pink or purple tassels at the ends of branches, turning green as they stiffen. *S. declinata* has smaller, orange flowers that darken to red with age. *S. indica*, the Asoka Tree, has a dense, pyramidal shape and narrow shining leaflets up to 20 cm. long.

These are all forest trees and do best in partial shade, in moist but well-drained soil. Propagation is by air-layering (marcotting) or through seeds, though seeds do not always breed true.

1 **Saraca declinata**   3 **Saraca indica** Asoka Tree
2 **Saraca indica** Asoka Tree

Solanum macranthum
(*S. wrightii*)
Potato Tree

Botanical Family: *Solanaceae*

This is a small, evergreen tree which may grow up to 12 metres in height but is more often seen in gardens as a large shrub. A native of South America, it has large, deeply lobed leaves and an almost continuous display of flowers, each up to 6 cm. in diameter, which are dark mauve when they open and fade to white over the course of a few days. Clusters thus appear as a very attractive mixture of mauve and white. The globular, tomato-sized fruits, are green at first and yellow-orange when they ripen.

    The Potato Tree needs full sun, rich soil, and frequent manuring to bloom profusely. It should be pruned occasionally to create a bushy shape, but this will inhibit flowering.

4

Melia azedarach
Persian Lilac, Chinaberry

Botanical Family: *Meliaceae*

Believed to have originated in northern India, this fast-growing, medium-sized tree is now found in many tropical and sub-tropical gardens. It has attractive bipinnate or tripinnate leaves and frequent panicles of small, pale lavendar, scented flowers, which are followed by yellow berries. In sub-tropical areas, where it grows to a considerable height, it can be used as a shade tree along avenues, but in the tropics it is more frequently seen as an ornamental in small gardens. It is often planted in temple gardens in Sri Lanka and Thailand.

    Propagation is easy by means of seeds or woody cuttings.

4 **Solanum macranthum**
Potato Tree

5 **Melia azedarach**
Persian Lilac

5

Erythrina
Coral Tree

Botanical Family: *Leguminosae*

Erythrina is a genus containing some 100 trees and shrubs, most with spiny trunks and branches and a number that produce strikingly beautiful flowers. *E. variegata* (*E. indica*), one of the most commonly seen, grows up to 15 metres and has large three-parted leaves that may be solid green or boldly patterned with bright yellow; in regions with a pronounced dry season, these fall once a year and are followed by beautiful scarlet blossoms. There is also a less common white-flowering variety. *E. crista-galli*, a smaller tree native to Brazil, has coral-red flowers as well as leaves almost all the time; *E. fusca* is also free-flowering; and *E. corallodendrum*, known as the Coral Bean, produces bright red seeds which are strung into necklaces.

Erythrina likes full sun and well-drained soil and bloom best after a period of drought. Quite large cuttings root easily.

1 *Erythrina crista-galli*

1

2

4

2 *Erythrina variegata*   4 *Erythrina fusca*
3 *Erythrina fusca*

3

29

1

Plumeria

Frangipani

Botanical Family: *Apocynaceae*

Originating in the New World tropics, from southern Mexico to northern South America and the West Indies, *Plumeria* is now found in all tropical countries, often in new varieties unknown to its native habitat. It belongs to the same family as the Oleander, Periwinkle, and *Allamanda*, and was named after Charles Plumier (1646-1706), a French botanist who made several voyages to the Caribbean area in the 17th century. (The genus name was originally spelled *Plumiera*).

The decorative tree, which ranges in size from small to several metres, has acquired an unusually large number of popular names: Dead Man's Finger (Australia), Jasmine de Cayenne (Brazil), Pagoda or Temple Tree (India), Egg Flower (Southern China), and Amapola (Venezuela). Two explanations have been given for the most common name, Frangipani: according to one it was inspired by an Italian perfume, created several centuries before the discovery of the Western Hemisphere; the other claims it was because the thick white sap that flows when the tree is cut reminded French settlers in the West Indies of "frangipanier", or coagulated milk.

It has an equally rich variety of cultural associations. The fragrant, five-petalled flowers are used by both Buddhists and Hindus as temple offerings and, in Hawaii, to make ceremonial leis presented on special occasions. In Malaysia and Indonesia the tree is often planted in Muslim cemeteries and is also a feature of most Buddhist monastery gardens. Traditional Thais are reluctant to use it in private gardens, not for its religious associations but because the local name, *lantom*, is similar to the word for "sorrow", though this superstition seems to be fading.

A 1938 study listed seven distinct species of *Plumeria*, though a more recent one

2

3

4

reduces the number to four. The most widely cultivated is *P. rubra*, from which countless popular garden varieties have been created, with flowers that may be white, pink, yellow, red, or in "rainbow" combinations. *P. obtusa*, sometimes called Singapore Plumeria, with intensely fragrant white flowers, has also been extensively hybridized and includes dwarf, shrubby forms as well as large trees.

Characteristically, *Plumeria* is multi-branched and has dense foliage that occurs at the ends of the branches; young wood is green and soft, later turning grey and harder. These are among the easiest of tropical trees to propagate, especially from cuttings, which no doubt explains their rapid spread on the ships of early explorers. Even a sizeable, well-branched cutting up to 2 metres long will root readily, though smaller tip cuttings are generally used. In a garden, the tree prefers full sun and good drainage but will grow in semi-shade. Except for *P. obtusa* and some of its varieties, which are mainly evergreen, most species of *Plumeria* have a dormant period during which growth stops, flowering is less profuse and the leaves tend to fall.

1 ***Plumeria rubra***	4 ***Plumeria rubra***
2 ***Plumeria rubra***	5 ***Plumeria rubra***
3 ***Plumeria obtusa***	'**Acutifolia**'

5

2

Thevetia peruviana

Yellow Oleander

Botanical Family: *Apocynaceae*

Closely related to the Oleander, with a similar growth pattern, this is often seen in gardens as a large shrub but in time becomes a small tree. It has shiny, lanceolate leaves that resemble those of the Oleander and regular displays of funnel-shaped, bright yellow flowers; there is also a form with apricot-coloured blooms. The flowers are followed by small green fruit, each containing two flattened seeds from which a heart medicine is extracted in some places. Both the fruit and the white latex that exudes from a cut in the bark are extremely poisonous.

Thevetia prefers dry, sandy soil and full sun. Seeds are slow to germinate and propagation is easiest by means of cuttings.

1 *Thevetia peruviana*	2 *Thevetia peruviana*
Yellow Oleander	Yellow Oleander

1

Mangifera indica

Mango

Botanical Family: *Anacardiaceae*

The mango is one of the oldest of cultivated fruits, mentioned in Indian writings some 4,000 years ago and common throughout Southeast Asia before the beginning of the Christian era. Numerous hybrids have been developed, ranging widely in fruit size, colour, and even taste, so that the mango familiar in, say, Mexico and Hawaii is very different from the one prized by Thais and Malaysians. Quite apart from its fruit, it makes an attractive shade tree for a good-sized garden, growing up to 30 metres with a dense, evergreen crown. Once or twice a year, particularly in regions with a pronounced dry season, small, greenish-yellow flowers appear, followed by the fruit.

For most profuse flowering, it needs full sun and good drainage. Propagation is usually by budding or grafting.

2

1

3

Fagraea fragrans

Botanical Family: *Loganiaceae*

In the opinion of some, this handsome tree is native to Singapore, where some beautiful specimens can be found in gardens and along roadways. It is slow-growing and remains bushy to the ground, with attractive glossy green leaves and, once or twice a year, a profusion of small fragrant flowers that are creamy white at first and turn yellow with age. These are followed by berry-like fruit that is attractive to bats and birds. The tree, though slow-growing, makes a useful tall screen in time or can be heavily pruned for use in smaller gardens.

It prefers at least partial sunlight and can be propagated by means of seeds or marcotting.

1 **Mangifera indica**
Fruiting Mango

2 **Mangifera indica**
Mango flowers

3 **Fagraea fragrans**

4 **Fagraea fragrans**

4

2

Pisonia grandis 'Alba'
Moonlight Tree, Lettuce Tree
Botanical Family: *Nyctaginaceae*

Relatively few books on tropical plants even mention this small tree, which is curious in view of how widely it is used as an ornamental, especially in the gardens of Southeast Asia. Its large, oblong-ovate leaves – pale yellow-green in the full sun, somewhat darker in shady locations – provide an effective foliage contrast in a predominantly green landscape and can also be a useful screen. It can grow to 3 metres or more in height, with a crooked trunk, but is more often seen as a large shrub. Flowers are rarely produced.

Pisonia does best in full sun or semi-shade. Propagation is by means of woody cuttings.

1 *Pisonia grandis* 'Alba'
Moonlight Tree

2 *Pisonia grandis* 'Alba'
Moonlight Tree

3

Spathodea campanulata
African Tulip Tree

Botanical Family: *Bignoniaceae*

A slender tree, which can grow up to 10 metres but is often kept smaller in gardens, this is a native of Uganda. It has dark green compound leaves and frequently produces clusters of large, tulip-shaped flowers at the ends of the branches. The most common flower colour is bright red-orange, with a yellow border on the petals; there is also a rarer, pure yellow variety. The blooms are followed by long, boat-shaped pods that open to release masses of winged seeds.

Spathodea grows rapidly from seed (those of the yellow-flowering variety do not, however, breed true) and prefers full sun, though it tolerates a wide range of conditions and poor soil. The wood tends to be brittle and, on taller specimens, breaks easily in strong winds.

3 **Spathodea campanulata**
African Tulip Tree

4 **Spathodea campanulata**
African Tulip Tree

4

1

Callistemon

Bottlebrush

Botanical Family: *Myrtaceae*

The popular name for this genus of twenty-odd shrubs and small trees from Australia derives from its distinctive bright red flowers which appear on most in cylindrical spikes at the ends of the branches. Some varieties can grow to a height of several metres but most are smaller. The narrow leaves are greyish-green on branches that may be upright or gracefully drooping in a way that suggests a willow tree. The branches continue to grow after the flowers have formed, so that new leaves appear at the tips.

Callistemon prefers full sun and a dry location; in moister gardens flowering is less regular and profuse. Smaller varieties can be grown as house plants in a sunny location.

1 *Callistemon viminalis*
Weeping Bottlebrush

2 *Callistemon viminalis*
Weeping Bottlebrush

2

Gliricidia sepium

Mexican Lilac

Botanical Family: *Leguminosae*

This native of tropical America can be rather straggly when young but is often planted in new tropical gardens because of its rapid growth. Older trees, especially if pruned carefully, can reach up to 10 metres in height and have an attractive shape. It has small leaflets, which drop during the dry season, and clusters of pale pink flowers along the twigs and bare branches. In South America it is known as "Madre de Cacao" since it is often used as a shade tree on coffee and cacao plantations.

Gliricidia prefers full sun and well-drained soil and tolerates severe pruning; it blooms best in areas with a pronounced dry season. Quite large cuttings, a metre or more in length, will root easily.

3

Schefflera actinophylla (*Brassaia actinophylla*)

Umbrella Tree, Octopus Tree

Botanical Family: *Araliaceae*

This native of Australia is a popular house plant in temperate countries because of its attractive, large, palmately arranged leaves, which resemble the spokes of an umbrella. In a tropical garden, it can become quite a large tree, 3 to 5 metres in height, producing striking inflorescences which appear from a central point in eight spikes, somewhat like octopus tentacles. The flowers are pale green at first, then become pink and finally dark red.

The Umbrella Tree prefers partial sun and well-drained soil. Since cuttings root slowly, air-layering is a more effective means of propagation.

3 **Gliricidia sepium**
Mexican lilac

4 **Schefflera actinophylla**
Umbrella tree

4

1 *Bauhinia purpurea*   3 *Bauhinia tomentosa*
2 *Bauhinia purpurea*

2

3

1

Bauhinia
Orchid Tree

Botanical Family: *Leguminosae*

Bauhinia is a large genus of some 300 species, among them vines, shrubs, and small trees. Perhaps the best known of the latter is *B. × blakeana*, popularly called the Hong Kong Orchid Tree, which has the two-lobed or twin leaves generally characteristic of *Bauhinia* and large, scented five-petalled purple flowers that do indeed resemble orchids. On the closely related *B. variegata*, the flowers range in colour from pure white through lavendar, on *B. purpurea* they are rose-mauve and on *B. monandra* pale pink.

Bauhinia grows quickly from seed, prefers full sun, and flowers best in areas where there is a distinct seasonal change.

Occasional hard pruning is desirable to make a bushier shape.

4 *Bauhinia tomentosa*

6 *Bauhinia acuminata*

5 *Bauhinia variegata* '**Candida**'
White Orchid Tree

5

1

2

Jacaranda mimosifolia
Jacaranda Tree

Botanical Family: *Bignoniaceae*

This bushy tree, growing up to 10 metres in height, is a native of Brazil and is prized by tropical gardeners because of its delicate fern-like foliage, smooth grey bark, and seasonal displays of beautiful tubular, blue-mauve flowers. There is also a less common, white-flowering form. It prefers areas with a distinct dry season, during which the leaves fall just before the flowers appear; in wetter parts of the tropics like Singapore and southern Malaysia, flowering is less regular and profuse, though the foliage remains attractive.

Jacarandas like sun and very well-drained soil. They are easily propagated by means of seeds or cuttings.

1 *Jacaranda mimosifolia*   2 *Jacaranda mimosifolia*

3

Cochlospermum

Buttercup Tree

Botanical Family: *Cochlospermaceae*

Once a year in the dry season, this medium-sized tree from tropical America sheds its deeply lobed leaves and puts forth a splendid display of large flowers, which resemble a camellia in shape and have bright gold petals and orange stamens. The flowers, either single or double, continue to appear for several months and fall to create a dazzling carpet around the base of the tree. The two commonly grown varieties are *C. vitifolium*, which grows to 6 or more metres in height, and *C. religiosum*, smaller but otherwise similar in form and flowering.

Cochlospermum requires full sun but is tolerant of most conditions, from dry to moist, providing there is good drainage. Some pruning after flowering helps to maintain a better shape. Propagation is by seeds or cuttings.

3 **Cochlospermum religiosum**

4 **Cochlospermum religiosum**

4

Scaevola taccada
(S. sericea)

Sea Lettuce, Beach Plum

Botanical Family: *Goodeniaceae*

Native to the coasts of the Pacific and Indian Oceans, this large shrub or small tree is a useful addition to seaside gardens since it tolerates very dry conditions and strong, salty winds. It is freely branching, often sprawling, and has fleshy, pale green leaves. The small, five-lobed white flowers appear on only one side of the corolla tube, leading to another popular name, "half flower". Its oval berries float for long periods, giving widespread distribution. *Scaevola* likes full sun and sandy soil but tolerates other conditions. It is propagated by seeds or by terminal cuttings.

1 ***Scaevola taccada*** Sea Lettuce

1

Terminalia catappa

Sea Almond

Botanical Family: *Combretaceae*

Tolerant of strong, salty winds, this is often seen growing along tropical coasts of the Indo-Pacific region and is a useful large shade tree for seaside gardens. It has very large green elliptical leaves, which in some places turn a bright red or yellow before falling twice a year, and grows in a distinctive layered form to a considerable height. The edible, almond-shaped nuts will float for long periods like the coconut, which is no doubt responsible for the tree's widespread distribution; bats are also attracted to the tannin-rich shell of the fruit. Seedlings grow rapidly and by the time of the first leaf-fall, after about four years, the tree is quite large.

2 *Terminalia catappa* Sea Almond

2

2

Samanea saman
Rain Tree, Monkeypod
Botanical Family: *Leguminosae*

Native to tropical America, the Rain Tree (or Monkeypod, as it is known in Hawaii and some other places) has spread throughout the tropics and for many years was the most popular garden shade tree. Because of its size, it is less often used in today's mostly small private gardens but is still seen along streets in cities like Bangkok and Singapore. It grows to a height of up to 24 metres, has bipinnate leaves, and when mature forms a characteristic flat-topped crown. The flowers, a mass of pink stamens, are followed by dark brown seed pods. Large Rain Trees are excellent hosts for epiphytic plants and the leaves are good for compost.

The popular name derives from the fact that the leaves close at night, or during heavy rain, which allows dew or raindrops to drip from them.

1 *Samanea saman* Rain Tree, yellow variety

2 *Samanea saman* Rain Tree

1

Delonix regia

Flame Tree, Flamboyant, Royal Poinciana

Botanical Family: *Leguminosae*

This large tree, which naturally assumes a spreading umbrella shape, originated in Madagascar and is now found in gardens, parks, and planted along streets throughout the tropics. It can grow as tall as 18 metres and has smooth greyish-coloured bark and attractive feathery, fern-like leaves, which drop annually during the dry season. At the same time as the new leaves appear, the tree is covered with masses of five-petalled flowers that are usually bright red-orange but may also appear in pale apricot. Blooming is followed by long, flattened, leathery dark brown or black seed pods.

Delonix requires full sun and adequate space to take its characteristic form. It flowers best in areas that have two distinct seasons but can be impressive even in moister places.

2

3

Peltophorum pterocarpum
Yellow Poinciana, Copperpod

Botanical Family: *Leguminosae*

Often seen as a shade tree in gardens and along avenues, this native of coastal Malaysia may achieve a height of 24 metres and forms a slightly flattened crown similar to that of *Delonix regia*. It has dark green bipinnate leaves and terminal branched clusters of bright yellow flowers, which are followed by coppery-coloured seed pods. In areas where there is a pronounced dry season, the leaves fall annually but reappear within a week or so along with the flowers.

While not as spectacular as *Delonix* in full bloom, *Peltophorum* offers better shade and is very ornamental in a sizeable garden. Seeds germinate slowly but grow fast once they have sprouted.

1 ***Delonix regia*** Flame Tree	3 ***Peltophorum pterocarpum*** Yellow Poinciana
2 ***Delonix regia*** Flame Tree	4 ***Peltophorum pterocarpum*** Yellow Poinciana

4

Artocarpus heterophyllus
Jackfruit

Botanical Family: *Moraceae*

This is a very stately tree, growing to about 20 metres, with glossy, dark green leaves. It is native to south India but is now grown over much of the tropical world. The huge fruit, yellowish-brown in colour and covered with small spines, appears almost continuously from the trunk and main branches. A single fruit can be nearly a metre long and contains edible seeds embedded in firm but

1

succulent sweet pulp. Like its relative the Breadfruit, it makes an attractive tree in large gardens; in Thailand, where its local name suggests "support", the tree is traditionally grown behind a house.

The Jackfruit prefers full sun and very good drainage. Propagation of desired varieties is by grafting.

Artocarpus altilis
Breadfruit

Botanical Family: *Moraceae*

In its native Polynesia, this large evergreen tree is grown primarily for its nutritious fruit – Captain Bligh was carrying Breadfruit seedlings on the *Bounty* from the Pacific to the West Indies, where it was planned to use them for slave food, when the famous mutiny erupted. Elsewhere, however, the fruit is less popular and the tree is grown more as an ornamental because of its large, handsome, dark green leaves, which have prominent yellow veins. It grows to a height of some 15 metres and takes a spreading shape, making it a good shade tree for larger gardens. The yellow-brown fruit appears at the ends of the branches. (A close relative is *Artocarpus heterophyllus*, the Jackfruit.)

Breadfruit trees need full sun and well-drained soil, but will grow under a light, high canopy.

2

1 **Artocarpus heterophyllus**
Jackfruit

2 **Fruit of Artocarpus**
Breadfruit

3 **Artocarpus altilis**
Breadfruit Tree

2

Casuarina

Botanical Family: *Casuarinaceae*

The most common member of this family is
C. equisetifolia, a tall tree often seen growing
20 metres or more in height along seacoasts
and also used as a light screen in gardens.
What appear to be the leaves are actually
small needle-like branches, which soon drop
and form a thick layer at the base. The trees
can be topped and clipped regularly to
make a bushy, semi-formal hedge; young
specimens have a conical shape, which is
lost as they grow and develop side branches.

Two other *Casuarina* species grown in
gardens are *C. nobilis*, which is a smaller,
denser tree, and *C. rumphiana*, on which the
needle-branches are finer and the branches
tend to droop attractively.

1 1 *Casuarina equisetifolia*   2 *Casuarina equisetifolia*

1

Gustavia superba ▲

Botanical Family: *Lecythidaceae*

This small dense tree from tropical America can reach a height of around 4 metres and a width of around 3 metres. It has large lanceolate leaves that are pinkish coloured when they first appear but later turn a glossy dark green. The striking creamy white flowers, which resemble lotus blossoms, are very large, up to 15 cm. in diameter, and strongly scented. These are followed by hemispherical, flat-topped fruit, each containing several large seeds.

Gustavia needs little pruning to maintain its bushy shape and makes a useful screen against low buildings. Seeds germinate easily.

Michelia ▼

Champaca

Botanical Family: *Magnoliaceae*

A member of the same family as the Magnolia, *Michelia* is widely grown in Southeast Asia, where its flowers are commonly used as temple offerings and in floral decorations. *M. champaca* is a small tree with glossy, light green leaves and flowers that range from yellow to orange. *M. alba* is considerably larger, growing up to 20 metres in height, and has creamy white blooms. The flowers, which have long, narrow, waxy petals around a greenish pistil, are not particularly striking in appearance but have a powerful fragrance that lasts long after they have been picked.

Michelia likes a sandy soil not subject to flooding, Propagation is by air-layering or cuttings, which root slowly.

2

3

Cananga odorata ▲

Botanical Family: *Annonaceae*

This is a medium-sized evergreen tree, around 10 to 15 metres in height, with slightly drooping branches. It is often grown in the gardens of Southeast Asia because of its flowers, which appear almost continuously on the leafy twigs. Though not striking from a distance, green when they first appear and becoming yellow with maturity, these are very fragrant and are used as offerings or to scent a room. There is also a smaller form that grows into a dense shrub about 3 metres in height.

Cananga does best in full sun but will grow in light shade. Propagation is by means of cuttings.

1 *Gustavia superba*   2 *Michelia alba*

3 *Cananga odorata*   4 *Coccoloba uvifera*
                          5 *Hibiscus tiliaceus*

Coccoloba uvifera ▼
Sea Grape

Botanical Family: *Polygonaceae*

Native to tropical and sub-tropical America, the Sea Grape is now found growing wild along many seashores and is also used as an ornamental in gardens subjected to salty winds and drought. It has stiff, rounded platter-like leaves shading from yellowish to olive green with prominent veins. The flowers are inconspicuous but the long clusters of berry-like fruit which follow are striking and give the tree its popular name. Somewhat acid in taste, the fruit can be used to make jelly. Other members of the genus have even larger leaves and make attractive ornamental plants when grown in a glasshouse.

Coccoloba prefers full sun and well-drained soil. Pruning improves the shape when it is used in a garden landscape.

5

Hibiscus tiliaceus ▲
Sea Hibiscus

Botanical Family: *Malvaceae*

Those familiar with the flowering *Hibiscus* shrub may not immediately recognize this small tree as belonging to the same genus. Native to the coastal regions of the Pacific and tropical Asia, it has large heart-shaped leaves which are slightly hairy on the undersurface and large five-petalled *Hibiscus*-type flowers that are bright yellow with a dark red base; as they age, the flowers darken and are orange by the time they fall in the evening. The tree tends to sprawl rather than grow upright but in time can develop quite a large trunk.

The Sea Hibiscus grows both along streams and in sandy areas exposed to salty winds, making it useful in seaside gardens where common *Hibiscus* is more difficult. Propagation is by seed.

4

1

Cerbera odollam
Pong-pong
Botanical Family: *Apocynaceae*

This hardy native of Malaysia, which grows
to a height of about 10 metres, is often used
as an ornamental in regional gardens since
it has attractive, glossy green leaves and
almost continuous displays of fragrant
white flowers with yellow centres. It can
be used for screening or kept pruned into
a small, dense tree. The flowers are followed
by round green fruits which are, however,
poisonous like those of other members of
the *Apocynaceae* family. *C. manghas* is a
closely related species which grows wild
on arid coasts.

Cerbera prefers a sunny location and
can be propagated by seeds or cuttings.
Flowering begins when the tree is
quite young.

2

3

Kigelia pinnata ▲
Sausage Tree
Botanical Family: *Bignoniaceae*

A native of tropical West Africa, this
medium-to-large tree – up to 15 metres
in height – is grown in gardens less for its
beauty than for its novelty value. It
produces large, dark red flowers that
hang from the crown of the tree and have
a rather unpleasant odour. When they fall
after a single night they are followed by
sausage-like fruit that may be as long as
50 cm., weighing several kilograms, quite
dramatic in overall effect.

Kigelia is a forest tree and will grow in
either sun or shade but requires a good
deal of space and so may not be suitable
for small gardens.

1 *Cerbera odollam*   3 *Kigelia pinnata*
2 *Cerbera odollam*   Sausage Tree

4

Couroupita guianensis
Cannonball Tree

Botanical Family: *Lecythidaceae*

Like the Sausage Tree, this native of South America is popular in botanic gardens and a few larger private gardens because of its bizarre fruit. It is a very tall tree which sheds its large, elliptical leaves several times a year. The pinkish-red flowers grow directly from the trunk near the base of the crown and are strongly scented in the evening. They are followed by large, globular fruit, reddish-brown in colour, which do indeed resemble cannonballs and remain on the tree for many months. On older trees there may be dozens of fruits and flower clusters at the same time.

4 *Couroupita guianensis*
Cannonball Tree

5 *Couroupita guianensis*
Cannonball Tree

5

2 *Flowering Shrubs and Annuals*

Flowering shrubs are among the most memorable features of any tropical garden. Though they may seldom fill the air with fragrance, particularly during daylight hours, they nevertheless offer a dazzling range of colour and flower forms.

Many of the more popular shrubs have been extensively hybridized to produce a constantly expanding choice of varieties. *Hibiscus,* for example, almost emblematic of the tropics to most people, can be found in countless forms, from small blossoms that hang like delicate pendants to huge ones the size of dinner plates. Equally varied are the popular *Ixora,* now available in large and dwarf species, and the showy *Mussaenda,* on which the actual flowers are unprepossessing but are surrounded by splendidly prominent white, pink, or red sepals. Nor is fragrance totally lacking: when evening falls, the hardly noticeable flowers of *Cestrum nocturnum,* Queen of the Night, may spread their perfume over an extensive area, blending with that of *Gardenia jasminoides,* a favourite in Southeast Asian gardens.

In a garden, flowering shrubs usually require full or at least a half day's sun to bloom well. Most also can be grown as exotic specimen plants under greenhouse conditions in temperate climates.

Left   **Mixed hedge of red-blooming *Russelia equisetiformis* and yellow *Allamanda* in an American tropical garden**

Above   ***Ochna serrulata***

1

3

Mussaenda

Botanical Family: *Rubiaceae*

Native to tropical Asia and Africa, *Mussaenda* is a small to large shrub with medium-sized oval leaves and prominent displays of colourful sepals surrounding small flowers. *M. philippica* 'Aurorae', one of the most popular in gardens, has pink, pink-and-white, or entirely white sepals and can grow up to 2 metres tall. *M. erythrophylla*, from West Africa, is a scandent shrub with bright red, velvety sepals resembling those of a Poinsettia; on *M. erythrophylla* 'Queen Sirikit' (named after Thailand's Queen), the sepals are light pink with a darker margin. *M. glabra* has a single, creamy white sepal instead of a cluster.

Mussaenda needs full sun to flower continuously and regular pruning to make a bushy plant. Propagation is from semi-hard cuttings.

2

1 **Mussaenda philippica**
2 **Mussaenda erythrophylla**
3 **Mussaenda philippica 'Aurorae'**

4

4 *Mussaenda philippica* 'Aurorae'

5 *Mussaenda philippica* 'Aurorae'

5

1

2

3

2 *Euphorbia millii*
'**Splendens**' Crown of
Thorns

3 **Euphorbia millii**

4 **Euphorbia pulcherrima**
Poinsettia

Euphorbia

Botanical Family: *Euphorbiaceae*

Euphorbia is a huge genus, comprising some 1,600 species that vary widely in appearance and habit but share the characteristic of producing a poisonous milky-white sap. Of the several grown as ornamentals, perhaps the best known is *E. pulcherrima*, the Poinsettia, a Mexican native on which the small flowers are surrounded by large, showy bracts. Blooming is conditioned by the amount of daylight the plant receives, which in many parts of the world accounts for the familiar displays at Christmas time. In addition to vivid red, the most common colour, there are also white and pink varieties, as well as cultivars with double bracts. Propagation is from cuttings.

Another popular *Euphorbia* is *E. millii* 'Splendens', the Crown of Thorns, a small shrub from Madagascar with thick, grey, spiny stems and clusters of small, bright coloured flowers. There are numerous cultivars, often grown as house plants or in rockeries.

1 *Euphorbia leucocephala* **with double and single forms of** *Euphorbia pulcherrima*

4

2

Lantana camara

Lantana

Botanical Family: *Verbenaceae*

Though *Lantana* is native to South America, it was an early introduction elsewhere in the tropical world, where it soon escaped the confines of gardens and became a weed. *L. camara* is a prickly shrub, about 1 metre in height, with ovate rough leaves and almost continuous displays of blooms that appear as clusters made up of tiny florets. Orange or red-orange are the commonest colours, but there are cultivars with larger white, pink, or lemon-yellow flowers. *L. sellowiana* is a trailing variety with purple flowers, often used as a ground cover.

Some people object to having *Lantana* in a garden because of the pungent smell of the leaves, but others appreciate the reliable display of bright flowers. In Thailand, several colours are sometimes grafted onto a single trunk and the result trained into a standard. *Lantana* always prefers full sun and well-drained soil. Propagation is by air-layering or from woody cuttings.

1 *Lantana camara*   2 *Lantana camara*

3

Plumbago auriculata (*P. capensis*)

Botanical Family: *Plumbaginaceae*

A native of the Cape Province of South Africa, *Plumbago* is popular in both tropical and sub-tropical gardens because of its delicate blue flowers; there is also a white-flowering cultivar. It is by nature a woody climbing shrub – with support it can reach several metres in height – but in the tropics it is seen more often as a low plant used for hedges. It tolerates both drought and poor soil and does best at slightly higher altitudes; heavy rainfall tends to beat down the flowers and leaves.

Plumbago requires full sun and grows best if it is heavily pruned from time to time, even cut back to the ground. Propagation is from green cuttings or by division of older plants.

3 *Plumbago auriculata*

1

Pereskia

Botanical Family: *Cactaceae*

Pereskia, native to tropical America, is a member of the Cactus family and therefore has spiny stems. One of the most attractive for use in gardens is *P. bleo* (*P. corrugata*), which can grow to 3 metres in height but can also be kept pruned into a bushy shrub; it has woody stems, fleshy, pale green leaves, and large orange or orange-red flowers that resemble roses. One variety, sometimes called the Wax Rose, has pink flowers. *P. aculenta*, more sprawling in nature, has creamy yellow to pinkish flowers and *P. grandifolia*, the "Rose Cactus", has dark green leaves and clusters of pink flowers.

    Pereskia likes full sun and dry conditions. Propagation is by means of cuttings.

2

Adenium obesum

Desert Rose

Botanical Family: *Apocynaceae*

Growing to a height of about 1 metre, this succulent plant is native to arid Arabia and East Africa and a member of the same family as *Plumeria*. Its swollen, often twisted trunk is pale grey, the leaves are glossy and club-shaped, and the flowers, which appear almost continuously, are trumpet-shaped and range from bright pink to crimson. It exudes a highly toxic sap which in some places is used as a poison for arrows.

    Adenium is not generally grown in moist tropical gardens but is often seen as a decorative pot plant; it may also be used in rock gardens. It needs full sun and a well-drained potting mixture.

1 **Pereskia bleo**   2 **Adenium obesum**
   (**Pereskia corrugata**)   Desert Rose

Thunbergia erecta

Botanical Family: *Acanthaceae*

Several of the best-known *Thunbergia* species are climbers, but this native of tropical Africa is a sprawling shrub, very useful in mixed beds or as a low hedge. It has attractive, small, dark green leaves and tubular flowers which in the most common form are rich purple with a yellow throat; there is also a white-flowering form, which seems to have smaller leaves and is somewhat less robust. *T. erecta* responds well to regular pruning and can be shaped into a very bushy shrub, though this inhibits blooming.

   It prefers full sun and well-drained soil but will bloom in partial shade. Propagation is by means of cuttings.

3

Wrightia religiosa

Botanical Family: *Apocynaceae*

This is a medium-sized shrub, up to 2 metres tall, from Thailand and Malaysia, with slightly pendulous branches, small oval leaves, and frequent clusters of small, fragrant white flowers. In Thailand, it is often planted in the gardens of Buddhist temples and also clipped into topiary shapes; it makes an attractive potted plant as well, requiring little in the way of care beyond occasional pruning for shape and applications of liquid manure.

   Wrightia prefers full sun but will also grow in light shade. Cuttings are slow to root and air-layering is a better means of propagation.

3 *Thunbergia erecta*   4 *Wrightia religiosa*

4

Calotropis gigantea

Crown Flower

Botanical Family: *Asclepiadaceae*

A member of the Milkweed family, this large, rather sprawling shrub thrives under dry, harsh conditions and can be found growing wild along seashores in Southeast Asia and the Pacific. The oval, conspicuously veined leaves are pale green and the star-shaped waxy flowers are either white or pale lavender; from the centre of each flower rises a small, elegant "crown" holding the stamens. In Thailand and Hawaii, the long-lasting flowers are often strung into leis or used in flower arrangements. The bark also yields a fibre that can be woven. Propagation is by air-layering or from woody cuttings.

Calotropis is particularly useful in sandy, exposed seaside gardens where other decorative shrubs are hard to grow.

1 *Calotropis gigantea* Crown Flower

1

Cestrum nocturnum

Queen of the Night, Night Syringa

Botanical Family: *Solanaceae*

In appearance, this tall shrub from tropical America with long scandent branches and pale green leaves is not very decorative; it often tends to sprawl if it is not frequently pruned. Nor are the small greenish flowers it produces periodically very noticeable. After dark, however, the flowers release one of the most potent fragrances to be encountered in a tropical garden. Some find it almost overpowering, which is why it is often planted at a distance from screened porches and verandas.

Cestrum grows in both sun and shade, and with regular clipping can be trained into a good hedge. Propagation is by cuttings.

2 ***Cestrum nocturnum*** Queen of the Night

2

1

Bixa orellana

Lipstick Tree, Annato

Botanical Family: *Bixaceae*

This native of tropical South America is an erect shrub with large, glossy green, prominently veined leaves that can become a small tree several metres tall. It is grown as a garden ornamental less for its pale pink flowers, which last only one day, than for its highly decorative two-valved seed pods, covered with soft, deep red hairs. The seeds inside are surrounded by orange-red arils, which yield a dye called annato that can be used in small quantities to colour lipstick or foods like cheese and margarine; the dried seed pods are also effective in flower arrangements.

When used as a hedge or screen, *Bixa* should be pruned often to encourage bushy growth at the bottom of the trunk.

1 ***Bixa orellana***
Lipstick Tree flower

2 ***Bixa orellana***
Lipstick Tree seed pods

2

3

Murraya paniculata

Orange Jasmine, Mock Orange

Botanical Family: *Rutaceae*

A member of the Orange family and native to India, this is one of the most often used shrubs in tropical gardens, and for good reasons. It has attractive, glossy green foliage, tolerates most conditions, and regularly produces clusters of small but strongly scented white flowers, followed by small, bright red fruit. *Murraya* makes an excellent hedge or screening plant, becoming in time a small tree up to 4 metres tall, or it can be clipped into formal shapes, though this will reduce the number of flowers. In Southeast Asia, the most profuse flowers occur during the rainy season.

It prefers full sun but will grow in partial shade, in almost any kind of soil. It also does well as a pot plant in greenhouses.

3 **Murraya paniculata**
Mock Orange

4 **Murraya paniculata**
Mock Orange

4

1

Gomphrena globosa
Globe Amaranth, Bachelor's Button

Botanical Family: *Amaranthaceae*

This herbaceous annual is native to dry parts of the New World but is also grown extensively in Southeast Asian gardens. It is low-growing with light green leaves and regular displays of long-lasting flowers that are globular in shape; magenta is the most common colour, but there are also white and pale mauve varieties. In Thailand the flowers are embedded in mounds of damp sawdust to make traditional flower arrangements in intricate geometric patterns.

When used in a garden bed, the plants will have to be replaced occasionally as they begin to look untidy. Propagation is from seeds, which take root easily.

Galphimia ▶

Botanical Family: *Malpighiaceae*

This busy shrub, a close relation of *Malpighia coccigera* (Singapore Holly), has small, glossy green leaves and almost continuous displays of small, bright yellow flowers. Since it tolerates a variety of conditions, including drought and wind, it is a useful garden bedding plant in tropical gardens and also does well as a pot plant in greenhouses.

Galphimia prefers well-drained soil and needs full sun to flower profusely but will grow in partial shade. It should be pruned back occasionally after flowering to make a better shape. Propagation is by means of cuttings or seeds, which are produced in abundance.

1 **Gomphrena globosa**
Bachelor's Button

2 **Galphimia glauca**

2

Barleria cristata ▶

Botanical Family: *Acanthaceae*

This is a small, easy-to-grow shrub, about
1 metre in height, which can be used for
a garden hedge or tightly clipped into
geometrical shapes. On the most common
variety, the tubular flowers are varying
shades of mauve; there is also one with pure
white flowers and another with flowers
striped in mauve and white. *Barleria
involucrata* is similar to *B. cristata* except that
both leaves and blooms are larger and the
flower colour is blue-lilac. *B. prionitis* is
a spiny bush with orange flowers.

Barleria is easily propagated from cuttings
planted directly in the ground. It requires full
sun to flower profusely and occasional
pruning to promote bushy growth.

4

3

◀ *Pachystachys lutea*

Botanical Family: *Acanthaceae*

Closely related to the genus *Belperone*, this
decorative shrub from the American tropics
is sometimes popularly called the Golden
Shrimp Plant. It grows to around 60 cm.
in height, has narrow, shiny leaves, and
prominent erect floral spikes that are actually
closely arranged bright yellow bracts from
between which the white flowers appear
over a period of time.

Given full sun, the shrub flowers regularly
in a garden bed and is also a useful pot
plant for greenhouses in temperate countries.
Propagation is by means of cuttings.

3 *Pachystachys lutea*   4 *Barleria cristata*
Golden Shrimp Plant

Hibiscus

Botanical Family: *Malvaceae*

If any single plant had to be selected as emblematic of the tropics, it would very likely be *Hibiscus*, the flowers of which brighten gardens from Bali to Florida. The genus contains some 200 species, found in both temperate and tropical areas, but the most commonly grown is *H. rosa-sinensis*. Some (though not all) authorities believe this to be a native of China; in any event, it has now spread throughout the tropical world in a wide selection of hybrids that range in

2

3

1

4

5

Also a member of the family, though quite different in outward appearance from the above, is *H. mutabilis*, popularly known as the Changeable Rose, the Rose of Sharon, or the Cotton Rose. This is a vigorous, bushy shrub growing to 2 metres or more, with greyish-green leaves and large flowers, either single or double, that are pure white when they open in the morning and change during the course of the day to deep pink.

Most *Hibiscus* require full sun to flower well and benefit from occasional pruning, especially when used as a hedge. Some of the hybrids developed in Florida and Hawaii have proved difficult to grow in Southeast Asian gardens and are more often seen in that region as carefully tended potted specimens.

1	*Hibiscus mutabilis*	5	*Hibiscus* **'Snowqueen'**
2	*Hibiscus* **hybrid**		**foliage, red leaved form**
3	*H. rosa-sinensis* **hybrid**	6	*Hibiscus* **'Snowqueen'**
4	*Hibiscus rosa-sinensis*	7	*Hibiscus schizopetalus*
			Coral Hibiscus

flower size from dainty to enormous, both single and double, and that go under an equally large variety of evocative names, among them Blue Bayou, California Gold, Hula Girl, Toreador, Brilliant, and White Wings. (In Indonesia, the common red *Hibiscus* is known as the "shoe flower", a reference to the fact that a juice extracted from the petals was supposedly used by the Dutch colonials to darken their shoes.)

Generally, *H. rosa-sinensis* is a medium-sized shrub, but some varieties can grow as tall as 5 metres. The leaves may be ovate or lobed, smooth or hairy, green or variegated. Flower colours range from pure white through lemon yellow and pink to scarlet. Though the flowers last only one day, many varieties bloom profusely so that there are nearly always several open at any time. Leaves and flowers are both edible and are sometimes used in traditional medicine.

Another popular species is *Hibiscus schizopetalus*, a native of East Africa, which has less dense foliage, arching branches, pendulous flowers with fringed petals that curl back against the stem and an exceptionally long staminal column. This is known in some places as the Coral Hibiscus, in others as the Fringed Hibiscus. Like

H. rosa-sinensis it has been used as the parent plant for numerous hybrids.

H. arnottianus is native to Hawaii and grows wild at higher elevations on the island of Oahu, becoming a large shrub or small tree; it has white flowers, usually with a red stamen, and unlike most *Hibiscus* is scented.

6

7

Nerium oleander
Oleander
Botanical Family: *Apocynaceae*

This is a tall, erect shrub with stiff, lance-shaped leaves and frequent clusters of white, pink, purple, or red flowers that may be single or double. A native of the Mediterranean area, it will grow in sub-tropical conditions and is often seen in gardens of the American South. It is notably resistant to salty sea winds and prolonged drought and it generally prefers dry conditions, which is one reason it is used less often in wet regions like Singapore and Malaysia. Like most members of the *Apocynaceae* family, all flowers of the plant are poisonous; the smoke from burning clippings will produce a reaction in some people.

2

Oleanders need full sun for good flowering and pruning to prevent them from becoming leggy. In areas of frequent rainfall they must be planted in very well-drained soil. They are propagated by cuttings or by air-layering.

3

Brunfelsia panciflora ▲
Yesterday, Today, and Tomorrow
Botanical Family: *Solanaceae*

The unusual popular name of this medium-sized shrub from South America becomes clear to anyone who observes it over the course of two or three days. When its fragrant tubular flowers first open they are a rich lavendar blue; then they change to pale lavendar and finally to almost white before they fall. Often all three colours can be seen on the same plant. It grows to about 1 metre and does best in filtered sunlight.

Another species, *B. americana*, larger than *B. panciflora* and also faster-growing, has creamy white flowers with long tubes, which are extremely fragrant after dark. In some places it is known as Lady of the Night, though a number of other scented plants have also been given this name.

Both species are propagated by cuttings.

1

1 **Nerium oleander**   3 **Brunfelsia panciflora**
2 **Nerium oleander**

4

5

Clerodendrum

Botanical Family: *Verbenaceae*

Clerodendrum is a large genus with nearly 400 species that include trees, shrubs, and climbers, several of them frequently seen in tropical gardens. *C. paniculatum*, native to Southeast Asia, is popularly known as the Pagoda Flower because its red-orange flowers appear in a conical form around a spike that rises high above the glossy five-lobed leaves. It may grow to 3 metres but is usually kept pruned to a lower height. Other species include *C. philippinum*, a small shrub which has fragrant white flowers tinged with pink that appear in clusters like an old-fashioned nosegay, and *C. ugandense*, the "Blue Butterfly", a scandent shrub on which the flowers are a mixture of pale and dark blue.

Propagation is by air-layering or from woody cuttings.

4 **C. quadriloculare**
5 **C. ugandense**

6 **C. paniculatum**
Pagoda Flower

6

Calliandra
Powder Puff

Botanical Family: *Leguminosae*

The name *Calliandra* means "beautiful stamens", and several of the approximately 150 shrubs and small trees belonging to this genus are grown for their dense heads of pink or white silky stamens that appear on short stalks along the upper sides of the branches. The leaves are pinnate with small fern-like leaflets that make the plant attractive even when it is not in flower. The most commonly seen species is probably *C. surinamensis*, which grows in a funnel shape and produces large numbers of pink-and-white or pure white flowers. A smaller species, with larger leaflets and bright red flowers, is *C. emarginata*; several other species also have red flowers.

Calliandra is propagated by cuttings or seeds.

1 *Calliandra surinamensis*   2 *Calliandra surinamensis*

2

1

Ixora

Botanical Family: *Rubiaceae*

Ixora (from the name of an Indian deity) is a genus with about 400 species native to India and tropical Africa, some of which rival *Hibiscus* as garden shrubs. *I. javanica*, one of the most often used, is a sizeable shrub with largish, pointed leaves and red-orange or pure red flowers that appear in rounded clusters of as many as sixty at the tips of the branches. *I. coccinea* is smaller, with glossy, more ovate leaves; usually red, but there are white, pink, yellow, and orange varieties, and a dwarf often used as a low hedge. *I. finlaysoniana*, a native of Thailand, can become a small tree and has large, fragrant, pure white flowers.

In full sun or light shade, *Ixora* flowers almost continuously needing little pruning except for a more formal shape. When grown as a house plant, it needs high humidity, warm temperatures, and exposure to strong light. Propagation is easiest by air-layering.

3 *Ixora javanica*   4 *Ixora coccinea*, **dwarf variety**

3

4

1

Duranta repens
Golden Dewdrop

Botanical Family: *Verbenaceae*

The popular name of this large erect shrub from tropical America is inspired by the clusters of bright orange-yellow berries that follow the flowers, in such quantities that they often cause the slender branches to droop gracefully. The lavender-blue or white flowers appear in loose clusters, and both berries and blossoms are often seen on a plant. There is also a form with variegated leaves, as well as one with bright yellow leaves; these do not bloom so profusely and are usually grown primarily for their foliage.

Duranta requires frequent pruning to make a bushy plant, especially when used as a hedge, and needs full sun to bloom well. The golden-leaf variety is often used as a ground cover or clipped into topiary. Propagation is by cuttings or from seeds.

2

1 *Russelia equisetiformis* (left) and a golden-leaved form of *Duranta repens* (right)

2 *Duranta repens* Golden Dewdrop

3 *Russelia equisetiformis* Coral Plant

3

Russelia equisetiformis ▲
Coral Plant, Firecracker Plant

Botanical Family: *Scrophulariaceae*

This is a low-growing shrub from Mexico, distinctive for its long, drooping, usually leafless stems and the almost continuous clusters of small, bright red tubular flowers that appear in racemes along them. Each plant produces many branches and is particularly effective when grown in a raised bed or rock garden so that the flower-laden stems hang down. Another species, *R. sarmentosa*, is a larger shrub with less showy flowers. Both varieties are attractive to butterflies and hummingbirds.

Russelia will grow under both moist and dry conditions, but requires full sunlight and occasional pruning to flower well. Propagation is by division of old plants or cuttings.

4

Caesalpinia pulcherrima
Dwarf Poinciana, Peacock Flower

Botanical Family: *Caesalpiniaceae*

Between 1.5 and 3 metres tall, this is one of the most popular tropical garden shrubs thanks to its rapid growth, ability to tolerate most soils, and almost continuous display of brightly coloured flowers that resemble in form those of its close relative, the Flame Tree or Royal Poinciana (*Delonix regia*). The flowers on the most common forms are red-orange or yellow, but other forms have pale yellow or cerise blooms. The shrub retains its ferny, light green leaves all year and benefits from hard pruning, which produces both a bushier plant and more profuse flowering. Black seed pods follow the flowers and should be removed, since otherwise they will remain on the plant for a long time.

4 *Caesalpinia pulcherrima*
Peacock Flower

5 *Caesalpinia pulcherrima*
Peacock Flower

5

Crinum

Spider Lily, Swamp Lily

Botanical Family: *Amaryllidaceae*

2

There are about 150 species in the genus *Crinum*, but only a few are cultivated as ornamentals. *C. asiaticum*, perhaps the most common, has a massive, fleshy stem and long, bladelike leaves; the fragrant, white, star-shaped flowers appear in large clusters on stalks that rise above the foliage. Native to Southeast Asia, it grows wild in both swamps and along the seashore, but is also often seen in gardens because of its attractive shape and tolerance of most conditions. The root is traditionally used as a poultice on wounds, while the heated and oiled leaves are applied to sprains.

Another species, *C. amabile* (sometimes listed as *C. augustum*), is larger and has pink flowers with red trimming. Unlike *Crinum asiaticum*, this requires full sun to bloom well. Propagation is achieved through separation of root suckers.

1

Hymenocallis littoralis ▷

Spider Lily

Botanical Family: *Amaryllidaceae*

Though called by the same popular name as *Crinum* and bearing somewhat similar white flowers, this is in fact a quite different plant. The long, strap-shaped leaves emerge straight from underground tubers and the star-shaped flowers have a membranous tissue connecting the petals at their base. There a number of cultivars, one with very narrow leaves, another with shorter, broader leaves and more compact flowers, and another with handsome, variegated foliage.

Hymenocallis is useful in a garden since it tolerates most soils, requires little attention, and will grow even in full shade, though some sun is preferred for good flowering.

1 **Crinum asiaticum**   2 **Crinum amabile**

3

5

3 **Hymenocallis littoralis**
Spider Lily

4 **Hymenocallis littoralis**
Spider Lily

5 **Hymenocallis littoralis**
'**Variegata**'

4

1

Jatropha multifida
Coral Plant

Botanical Family: *Euphorbiaceae*

This is a decorative shrub with sparse branches that grows up to 3 metres. The light green leaves are large and lobed, the flowers small but bright red or pink and appearing in considerable numbers on red stems. Another species, *J. podagrica*, is a smaller plant, about 60 cm. high, with a knobby, swollen stem and red flowers in large terminal clusters.

Both make good potted plants but can also be used in a small garden as border plants. Partial sun and a well-drained soil are preferred. Propagation is by seeds, which are produced freely and germinate easily, or by cuttings.

1 *Jatropha multifida*
Coral Plant

2 *Jatropha podagrica*

2

3

Hamelia patens ▽
Fire Bush

Botanical Family: *Rubiaceae*

This native of tropical America, sometimes called the Firecracker Plant, has pale green rounded leaves and small, tubular red-orange flowers. Though it can grow to 2 metres or more in height, it responds well to hard pruning and can be used to make a dense medium-sized hedge or an attractive shaped plant; pruning, however, inhibits the blooms, which appear at the ends of the branches.

Hamelia likes full sun and very well-drained soil. Propagation is easy by means of cuttings.

4 *Hamelia patens* Fire Bush

Dillenia suffruticosa ▲
(Wormia suffruticosa)

Botanical Family: *Dilleniaceae*

This is a hardy native of Southeast Asia, often seen growing wild and also used in low-maintenance gardens since it tolerates most conditions and needs little attention. It can grow up to 5 metres in height but is more often seen as a large shrub. The large, cabbage-like leaves are oval and have a slightly jagged edge; the flowers are large and yellow, followed by bright red fruit. The flowers last a single day but are produced in profusion so there are nearly always several in bloom.

Dillenia will grow in sun or shade but prefers a little light. Propagation is by means of seeds, though seedlings grow slowly at first.

3 *Dillenia suffruticosa*

4

1

2

3

Canna indica

Canna Lily

Botanical Family: *Cannaceae*

Despite its botanical name, *Canna indica* comes not from India but from tropical America and the Caribbean. It was introduced elsewhere quite early, however, and the numerous hybrids have long been a favourite with gardeners throughout the tropics, especially planted in large beds. The commonest flower colours are red and yellow, but they are also found in orange, pink, cream, and mixtures and with many leaf variations.

The fleshy stalks grow to around 1.5 metres and flower best in full sunlight; each stalk must be cut back to the ground after blooming. The underground rootstocks spread rapidly and in time may become overcrowded. Some gardeners recommend digging up the entire bed annually, dividing the roots, and replanting in freshly manured soil.

1 **Canna indica** Canna Lily   3 **Canna indica** Canna Lily
2 **Canna indica** Canna Lily

Impatiens

Balsam, Busy Lizzie

Botanical Family: *Balsaminaceae*

This is a large genus, some species of which are native to Southeast Asia and others to tropical Africa, with numerous ornamental cultivars that have become popular as pot plants or greenhouse specimens in temperate countries. *I. balsamina*, commonly known as the Annual Balsam, grows wild in Indonesia; it has a fleshy stem that grows to about 70 cm. and white, pink, red, or purple flowers. The Busy Lizzies are mostly cultivars of *I. walleriana* (sometimes listed as *I. holstii*) and have a wide range of flower colours, both single and double, and leaves that may be green, purple, or variegated.

In cooler parts of the tropics like Hawaii, Busy Lizzies can grow into quite sizeable shrubs and are used effectively as bedding plants. They do not grow as vigorously in hotter areas but can be cultivated in moist semi-shaded locations beside a pool or waterfall. Propagation is by cuttings or seeds, though the seeds of some hybrid varieties do not always breed true.

4

5

4 *Impatiens walleriana* **cultivars**   5 *Impatiens walleriana* **cultivars**

1

Pseuderanthemum reticulatum

Botanical Family: *Acanthaceae*

This a useful shrub in a garden since both its foliage and flowers are attractive. The lanceolate leaves are bright yellow when they first appear, later green with yellow veins, thus providing a striking colour contrast when planted in a mixed bed. It also produces frequent sprays of white flowers at the branch ends. Though it can grow to over 2 metres, it is usually kept shorter and bushier by pruning. Another species, *P. atropurpureum*, has deep purple leaves and similar, but less prominent, flowers.

Both species need full or partial sun for the leaf colour to be at its best. Propagation is by means of woody cuttings.

2

3

Eranthemum nervosum (*E. pulchellum*)

Blue Eranthemum

Botanical Family: *Acanthaceae*

This is a strongly branched shrub of Indian origin, popular with gardeners because of the spikes of small flowers it produces. These are often gentian blue – an unusual colour in the tropics. The flowers appear from green-and-white veined bracts that remain after the blooms fall, forming a column several inches long. The hairy leaves are large and dark green.

A sprawling shrub which may reach a metre or more in height, *Eranthemum* is usually kept lower and bushier through pruning. Light shade is preferred in a garden; in a greenhouse it needs warm conditions. It is easily propagated from cuttings.

1 *Pseuderanthemum reticulatum*

2 *Pseuderanthemum reticulatum*

3 *Eranthemum nervosum* Blue Eranthemum

4

Tecomaria capensis
Cape Honeysuckle

Botanical Family: *Bignoniaceae*

Sometimes now listed as *Tecomia capensis*, this is a sprawling plant that, if left unpruned, puts out long stems and becomes almost a vine that needs some kind of support. Usually kept clipped, it is often used as a low hedge. It has attractive, light green foliage and frequent displays of long, tubular flowers that may be yellow or bright orange at the ends of the stems. It spreads by means of underground roots that can become invasive.

A native of South Africa, Cape Honeysuckle needs very well-drained soil and prefers full sun, but will bloom in filtered light; flowering is best is slightly cooler climates. Propagation is by means of seed cuttings, or branches that root when they lie on the ground.

4 *Tecomaria capensis*
Cape Honeysuckle

5 *Tecomaria capensis*
Cape Honeysuckle

5

1

Ochna serrulata ▲

Mickey Mouse

Botanical Family: *Ochnaceae*

This is a useful ornamental shrub from
tropical Africa with glossy green leaves
and bright yellow flowers that grows to
around 2 metres in height. The popular
name derives from the small, jet-black
seeds that appear in pairs surrounded by
bright red sepals and suggest to some
people the face of the famous cartoon
character. A hardy plant, it tolerates most
soils and attracts few insect pests. It grows
best in full sun and needs some pruning
to develop a more attractive shape.

Propagation is from seeds or half-green
woody cuttings.

Gardenia jasminoides ▼

Gardenia

Botanical Family: *Rubiaceae*

Popular in tropical and sub-tropical
gardens and also used as a house plant
in temperate climates, this native of South
China is a medium-sized shrub with shiny
dark green leaves.

The pure white flowers, usually double,
are very strongly scented and appear more
or less continuously on a healthy plant in
the tropics.

Numerous cultivars have been developed,
including one with variegated leaves and
another that grows to only about 50 cm.
Another species, *C. carinata*, the Malaysian
Tree Gardenia, can become several
metres tall but flowers less profusely than
the shrub varieties.

Gardenias require full sun and occasional
spraying against mealy bugs and other
insect pests. Propagation is from woody
cuttings or by air-layering (marcotting).

3

Kopsia fruticosa

Botanical Family: *Apocynaceae*

Native to Southeast Asia, this is a slow-
growing shrub that can become quite tall
but is usually seen as a spreading bush
around 1 or 2 metres in height. It has pale
grey bark, light green, spear-shaped,
somewhat leathery leaves, and almost
continuous *Vinca*-like flowers, pale pink or
white with a crimson centre, which appear
at the ends of the branches. It can be
pruned into a bushier shrub but is most
attractive when left to assume its natural
open shape.

Kopsia needs at least a half day's sun to
bloom well. Propagation is by means of
cuttings or air-layering.

2

1 **Ochna serrulata**

2 **Gardenia jasminoides**
  Gardenia

3 **Kopsia fruticosa**

4 **Tabernaemontana
  divaricata**

5 **Tabernaemontana
  divaricata**

6 **Tabernaemontana**
  Paper Gardenia

Tabernaemontana

Botanical Family: *Apocynaceae*

This is a medium to large shrub with glossy
green elliptical leaves and snow-white
flowers that may appear single in a pinwheel
shape or double. It goes by a variety of
popular names such as Crape Jasmine and
Paper Gardenia and is fragrant at night,
though not as strongly as the true jasmine
or gardenia. The most popular species, used
for mixed beds or as a low hedge, is the
Pinwheel Flower, *T. divaricata*, which
blooms frequently throughout the year.

5

6

4

1

Brugmansia x candida (Datura candida)

Angel's Trumpet

Botanical Family: *Solanaceae*

Belonging to the Deadly Nightshade family, which also includes the tomato, *Brugmansia* comprises about 20 different species, nearly all of them poisonous. *B. × candida*, which in time can become a small tree, makes a striking garden ornamental with its large, white, trumpet-shaped flowers that appear in great quantities hanging like bells. The velvety, greyish-green leaves are spear-shaped and the wood is quite brittle. There are a number of cultivars, some with pink or purple flowers.

Believed to be native to the mountains of Chile and Peru, *B. × candida* grows better at higher elevations than in lowland gardens. Propagation is by seed or cuttings.

1 *Brugmansia x candida*

Medinilla

Botanical Family: *Melastomataceae*

This genus of about 300 species contains both erect and epiphytic shrubs. The best known in tropical gardens is *M. magnifica*, an impressive shrub native to the Philippines which grows up to 2 metres in height, has large, leathery leaves with an ivory-coloured midrib, and periodic masses of pink-red flowers surrounded by showy pink bracts. A common name for this species is the Rose Grape. Another species is *M. cumingii* , an epiphyte which produces very showy displays of pink or light red flowers that hang in pendulous clusters.

    Medinilla does not do well at lower altitudes and is therefore not often seen in the lowland gardens of Southeast Asia. It is often used in Hawaii, however, where temperatures are cooler and are also popular in greenhouses. Propagation is by air-layering or woody cuttings.

2 *Medinilla magnifica*

2

1

2

3

Catharanthus roseus ▲ *(Vinca rosea)*

Madagascar Periwinkle

Botanical Family: *Apocynaceae*

This perennial shrub, which some mistake for an annual, grows to a height of about 60 cm., has soft, hairless stems, and produces an almost continuous display of rosy pink or white five-lobed flowers. Most authorities believe it to be a native of Madagascar, though some cite the Caribbean as its place of origin; in any case, it is now found in most tropical countries, sometimes growing wild in dry, sandy areas and often cultivated in gardens because of its dependable flowering. Like other members of the *Apocynaceae* family, it is poisonous, but it is also used for medicinal purposes and in recent years has been the subject of considerable scientific research.

It prefers a sunny, well-drained location and is easily propagated from seeds.

Malvaviscus arboreus

Wax Mallow, Firecracker
Hibiscus, Turk's Cap

Botanical Family: *Malvaceae*

Very similar in appearance to *Hibiscus*, this bushy shrub from tropical America has hairy, toothed leaves and large flowers that are rolled as if about to open. The usual colour is bright red, but there is a pink-flowering cultivar. The shrub can grow up to 4 metres in height and can become invasive in a small garden unless regularly pruned back; though it is attractive to butterflies and hummingbirds, it is also often attacked in some areas by leaf-eating insect pests and may need regular spraying.

Malvaviscus needs full sun and a well-drained but moist soil. Propagation is by means of cuttings, which root easily.

1 *Malvaviscus arboreus* 3 *Catharanthus roseus*
2 *Malvaviscus arboreus*   Madagascar Periwinkle

Cordia sebestena
Geiger Tree

Botanical Family: *Boraginaceae*

Cordia is a genus of some 250 species, most of them native to tropical America. *C. sebestena*, which can grow up to 8 metres in height, is often listed as an ornamental tree though it is more usually seen in gardens as a large shrub. It has elliptical, rough-textured leaves and frequent clusters of bright orange tubular flowers that appear at the ends of stems. The flowers are followed by round, sweet fruits which in some places are used as a cough remedy. Another species, *C. boissiere*, is much larger and has creamy white flowers.

Cordia needs full sun and does well in dry conditions. Propagation is by means of cuttings or seeds.

4

Tecoma stans
Yellow Bells, Yellow Trumpet Tree

Botanical Family: *Bignoniaceae*

This tall, erect shrub from tropical America, which can become a small tree, regularly produces large clusters of bright yellow, trumpet-shaped flowers, which explains its popularity in many gardens. It has light green pinnate leaves and grows quite rapidly, developing within a few months into an attractive plant that is useful for screening or in mixed beds.

To make a bushier shrub and increase the number of blooms, the branches should be cut back after flowering. It can be propagated from seeds or, more rapidly, from cuttings, which root easily.

4 **Cordia sebestena**
Geiger Tree

5 **Tecoma stans**
Yellow Bells

5

3 *Foliage Plants*

Flowers are not the only source of colour in a tropical garden or greenhouse. They can be accompanied by numerous specimens on which the blooms are inconspicuous or relatively uninteresting but which offer, as if in compensation, a brilliant array of foliage hues and unusual leaf forms. Some of these were among the earliest exotics brought back to the West by European collectors, attracting crowds when they were first exhibited and later winning popularity as cherished house plants. In the tropics, often far from their countries of origin, they have been equally popular with gardeners seeking to create bright contrasts and variety amid the predominant green, particularly in places where flowering shrubs are difficult because of heavy rains or insufficient light.

Codiaeum, for instance, found in a vast variety of multicoloured hybrids, is among the favourites for sunnier areas, as is *Cordyline*, several species of *Dracaena*, and *Acalypha*. Moist, shady areas are ideal for *Aglaonema*, *Dieffenbachia*, and *Caladium*, members of the Aroid family celebrated for their ornately patterned leaves, as well as for the jewel-like designs of *Calathea* and *Maranta*. In some South American and Hawaiian gardens, almost all the colour in both sun and shade is provided by different species of Bromeliad, many of which have the added advantage of occasionally producing spectacular floral displays.

Left   **A mixed bed of foliage plants including *Dieffenbachia, Cordyline* and *Dracaena***

Above ***Dracaena marginata***

1

Begonia

Begonia

Botanical Family: *Begoniaceae*

Native to a number of tropical and subtropical parts of the world, Begonias have long been popular greenhouse plants in temperate countries and are also used in some gardens, especially those with slightly cooler temperatures. Many hybrids have been developed, some grown primarily for their ornamental foliage, others for their attractive flowers. Among the latter group, which come mostly from tropical American species, are *B.* × *semperflorens*, which has small pink or white flowers, both double and single, and *B. popenoei*, on which the panicles of white flowers rise upright above the leaves.

Among the foliage Begonias, *B. rex*, native to Assam, is one of the best known, along with *B.* × *erythrophylla*, which has dark green leaves that are purple and red underneath, and *B. heracleifolia*, with striking green-and-bronze star-shaped leaves.

Nearly all Begonias prefer moist but well-drained soil and shady conditions, though some will grow in filtered shade. Propagation is by stem or root cuttings or by division of rhizomes.

1 *Begonia dichotoma*

Calathea

Botanical Family: *Marantaceae*

Mostly grown in gardens for the handsomely patterned foliage, *Calathea* is native to Central and South America and the Caribbean. On *C. makoyana*, the Peacock Plant, the leaves are pale green with a darker green border on top and a blend of green and purple below; on *C. metallica* the undersides are silvery green and red and on *C. majestica* the top is patterned with white or pink stripes. One variety grown for its inflorescences is *C. lancifolia*, the Rattlesnake Plant, which produces tall stalks of flowers enclosed in yellow bracts.

Calathea prefers moist, shady, jungle-like conditions in a garden and is also often grown as potted specimens. When it is used as a house plant, the leaves should be sprayed with water several times a week. Propagation is by root division.

2 **Calathea cultivar**   3 **Calathea micans**

1

3

2

4

5

Acalypha
Copper Leaf, Beefsteak Plant
Botanical Family: *Euphorbiaceae*

A native of the East Indies and the Pacific, this is one of the most striking foliage shrubs and is widely used by tropical gardeners. Cultivars of *A. wilkesiana*, which can grow as tall as 2 metres, are found in a wide variety of colours: green and white, green and yellow, red, bronze, coppery, or brown. The leaves also vary in form, some being flat and others undulating, while the flowers are inconspicuous.

A. hispida has plain green leaves but compensates by producing masses of long, bright red inflorescences that hang down and give rise to its popular name of "Cat's Tails".

In a garden, where they are often used as informal hedges or in massed beds, all *Acalypha* species require full sun, well-drained soil, and careful pruning to prevent them from becoming leggy. They are easily propagated by cuttings.

1–4 *Acalypha wilkesiana* **cultivars**   5 *Acalypha hispida* Cat's Tails

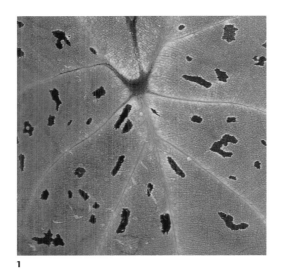

1

2

Caladium

Botanical Family: *Araceae*

This native of tropical South America and the Caribbean is perhaps the best-known of the ornamental Aroids, widely cultivated in both greenhouses and tropical gardens. *C. bicolor* was introduced to Europe from the New World by early Portuguese explorers and by the 18th century was a popular hothouse plant with numerous different leaf colourations, ranging from almost pure white to bright scarlet. Those with large leaves have been called *C. × hortulanum*, while others with smaller leaves were called *C. picturatum*; another with very small leaves is *C. humboldtii*.

In a garden, *Caladium* likes rich, well-drained soil and moist, cool conditions. It is likely to die back during a dry spell but will return with the rains and spread. The tuberous bulbs of potted specimens should be dried off and rested after completing their growth cycle. Propagation is by plant separation.

3

Anthurium

Botanical Family: *Araceae*

To non-tropical gardeners, *Anthurium* is probably best known in its flowering forms, which are treated in a different chapter. Several species, however, are grown for their ornamental foliage, usually in pots but, given proper conditions, sometimes in the ground. The best-known is *A. crystallinum*, a native of Colombia and Peru, which has huge, velvety, dark green leaves laced with dramatic white veins. Another popular species is *A. hookeri*, the Bird's Nest Anthurium, which grows in a rosette of large spatulate leaves that may become over 1 metre long and ½ metre wide. The flowers on both of these are relatively inconspicuous.

In either a garden bed or in a pot, *Anthurium* needs very well-drained soil containing a large amount of humus and protection from strong winds and sunlight. It can be propagated by plant separation.

4

4 *Anthurium podofillum*   5 *Anthurium crystallinum*

5

Dracaena

Botanical Family: *Agavaceae*

This is a large genus that includes plants of varying sizes and leaf colourations, many of which are used in tropical gardens. *D. fragrans*, one of the most commonly grown, has long strap-like leaves and can become a small tree; it occurs in a number of forms, some with green leaves and others on which the leaves have bold yellow stripes. Tolerant of neglect, it is often seen as a house plant or in garden areas that receive little attention.

D. reflexa (also known as *Pleomele reflexa* and, popularly, as Song of India) has small, narrow leaves which may be dark green or have bright yellow margins. Another useful species is *D. marginata*, sometimes listed as *D. concinna*, which has long flat leaves striped with red or purple; though it can grow quite tall, it is more often seen in lower masses to bring colour to a garden.

Dracaena prefers partial to full sunlight, though green leaved forms will do well in shady locations. All are easily propagated from stem cuttings.

1 *Dracaena reflexa*
Song of India

2

3

4

2 **Wild** *Dracaena*   4 *Dracaena fragrans*

3 *Dracaena surculosa*   5 *Dracaena marginata*

5

Bromeliads

Botanical Family: *Bromeliaceae*

The Bromeliads comprise a very large family of some 52 genera and 2,500 recognized species, a number that is constantly growing through hybridization. All but a single species, *Bromelia pitcairnia*, which comes from Senegal, are native to tropical South America and the Caribbean, are found in locations that range from desert to rain forest and are both terrestrial and epiphytic in growth. The family name is derived from *Bromelia*, one of the largest genera.

Christopher Columbus is supposed to have discovered the first cultivated Bromeliad when he landed on Guadaloupe on his second voyage to the New World. This proved to be the famous pineapple, *Ananas comosus*, which on sampling he found to be "delicious and freshing to the taste". Samples were brought back to Europe and eventually spread as a commercial crop to most parts of the tropical world in hybrid forms. Although pineapples are generally

seen in plantations rather than gardens, one cultivar, *A. comosus* 'Variegata', has handsome striped leaves with red spines along the margins and is often used as an ornamental.

Most Bromeliads are epiphytic, growing on trees, fallen branches, and rocks without soil. One of the largest and most colourful groups is *Aechmea*, which has broad, fleshy leaves in a rosette and striking inflorescences that rise from the centre of the plant. The best-known member of this genus is *Aechmea fasciata*, often grown as a potted specimen in green-houses. Other widely cultivated genera are *Billbergia*, *Bromelia*, and *Neoregelia*, most of which have coloured leaves or prominent flower stalks. The genus *Tillsandia* is found from the southern United States to northern Argentina and Chile; the leaves are usually

4

3

5

grey and spineless, the inflorescences often in startlingly brilliant colours. (One unusual *Tillsandia* is the so-called Spanish Moss which hangs profusely from so many trees in the American South.) The majority of Bromeliads are propagated by offshoots that develop from the base of the parent plant after flowering.

In Hawaii, Florida, and many parts of Central and South America, Bromeliads are used effectively in gardens, either grown on trees or rocks or in massed plantings. They are less common in Southeast Asian gardens, especially those at lower altitudes, though they are often seen as potted specimens. They are also popular as house plants in temperate countries.

1–3 **A massed planting of Bromeliads, with flower details**

4 *Ananas comosus* **'Variegata'**

5 *Bromelia pitcairnia*

2

Pandanus

Screw Pine

Botanical Family: *Pandanaceae*

Some members of this large genus of over 600 species, like *P. tectorius,* grow wild along the seashores of the tropical Pacific and can become small trees, while other, shrubbier species lend themselves to garden landscapes. The popular name derives from the fact that the long, prickly leaves emerge in a screw-like arrangement. Perhaps the most decorative species is *P. sanderi,* with green and yellow striped leaves. Similar in appearance, *P. veitchii* has green and white stripes. There is a cultivar of *P. sanderi* with spineless leaves. Both eventually become quite large, 6 to 8 metres in height, and so must be carefully sited in a garden.

Pandanus is a very adaptable plant, though it prefers hot, dry conditions. The easiest method of propagation is by cuttings, which root readily when placed directly in the soil.

1 *Pandanus tectorius*   3 *Pandanus utilis*
2 *Pandanus tectorius*   4 *Pandanus amaryllifolius*
                          5 *Pandanus sanderi*

1

3

4

5

109

Cordyline fruticosa

"Ti" Plant

Botanical Family: *Agavaceae*

Cordyline fruticosa (*C. terminalis*) popularly known by its Polynesian name "ti", occurs in such a vast number of colours and leaf forms that one may be forgiven for thinking of it as a range of species rather than just one. It was once classified botanically among the *Lilaceae* but has been moved to the *Agavaceae*, which also include the Agave, Yucca, and *Dracaena*.

The plant may have single or multiple stems, or canes, which support clusters of glossy leaves from the top. The flowers appear periodically in panicles from the leaf cluster and are followed by small seeds enclosed in berry-like fruit.

Native to Southeast Asia and parts of the tropical Pacific, *Cordyline* derives its name from the Greek *kordyle*, meaning a club, a reference to the large, club-like roots produced by some species; *fruticosa* means shrubby; (and *terminalis*, meaning end, refers to the fact that the flowers appear from the top of the leaf clusters). The original Polynesian settlers of Hawaii brought with them a green variety which was viewed as a symbol of divine power and used in leis worn by priests. At some later point, the red *Cordyline* was introduced, possibly from Melanesia, and for a long time these were the only colours grown in the islands. Aside from the colour, another important difference between the two is that whereas the green variety produces no seeds the red one does, meaning that the latter could be used to produce hybrids with leaves in new colours.

Extensive cultivation of *Cordyline* varieties for ornamental use in gardens as well as for house plants began in the early years of the 20th century, not only in Hawaii but also in Trinidad and, later, in Thailand. Through cross-pollinination, literally hundreds of varieties have resulted, displaying almost every colour in the rainbow, ranging in size from large to miniature, and bearing such descriptive names as Hawaiian Flag, Hilo Rainbow, Indian Blanket, Maui Beauty, Pink Diamond, and Lovely Hula Hands. In tropical countries with a distinct change of seasons, *Cordyline* colours become richer and more vivid during the cooler, drier months that follow the rains.

The easiest method of propagation is by tip cuttings that include about 6 inches of the cane or by short lengths of the cane placed either vertically or horizontally in the rooting medium. Seeds also germinate in 5 to 10 weeks, depending on the variety, though because of cross-pollination or genetic instability they may produce plants that differ from the parent.

In a garden, most varieties need sun to develop full leaf colour, though some prefer semi-shade. As house plants they also like a sunny location, warmth in the winter, and plentiful watering during the summer months.

1–4 *Cordyline fruticosa* cultivars

Alocasia

Botanical Family: *Araceae*

There are some 70 species of *Alocasia*, all native to tropical Asia and many with distinctive foliage. One of the largest is *A. macrorrhiza*, popularly known as the Elephant Ear, on which the leaves can be 1 or 2 metres long growing out of a thick trunk. In many places, these are grown for their starchy tubers but they also make dramatic garden plants, especially a variegated form with green-and-white patterned leaves. Other decorative *Alocasia*

2

species include *A. sanderiana*, with prominently lobed leaves patterned in silvery white; *A. cuprea*, with leaves the colour of burnished metal; and *A. longiloba*, glossy green on top and shiny purple below.

Alocasia is propagated by separating offshoots that appear from the parent plant. Generally it prefers damp, cool locations and protection from strong sunlight and winds.

1

3

Strobilanthes dyerianus ▲
Persian Shield

Botanical Family: *Acanthaceae*

This herbaceous shrub, a native of Burma, has beautiful long, narrow iridescent leaves that are purple with silver markings on top and purple beneath; new leaves have a strong silvery purple flush. Though it can be planted in a garden bed that receives filtered sunlight, it is more often seen as a potted specimen. In time, it tends to become leggy and should be pruned or pinched back when young to make a bushier, more attractive plant.

Propagation is by cuttings. In a greenhouse it prefers moist, warm conditions.

Philodendron ▼

Botanical Family: *Araceae*

Most popularly cultivated *Philodendron* species are climbers and are treated as such in a separate chapter. Some, however, are what is known as "self-heading", which means that the foliage appears as a crown at the top of a stout, thick trunk, which is supported by aerial roots as it grows taller. These are among the most "tropical" looking of all plants in a garden. One often seen is listed in some books as *P. bipinnatifidum*, in others as *P. selloum*; sometimes both are named, though authorities now regard them as the same species. The leaves become larger and more lobed as the plant grows, eventually reaching 1 metre or so in length on long stalks, and the whole plant can become about 2 metres tall. It prefers light to medium shade and moist, well-drained soil.

Since self-heading species of *Philodendron* have a single growing point, it is difficult to propagate them from cuttings; the easiest method is by seeds, which should be sprinkled on a fibrous medium and kept moist until they germinate.

1 ***Alocasia macrorrhiza***
  Elephant Ear

2 ***Alocasia sanderiana***

3 ***Strobilanthes dyerianus***

4 ***Philodendron***
  bipinnatifidum

4

1

Maranta

Botanical Family: *Marantaceae*

Often confused with *Calathea* by non-specialists, many species of *Maranta* have similarly patterned leaves. On *M. leuconeura*, a low, evergreen plant that grows to about 1 foot in height, the oval leaves are a dark blue-green with red or white veins; in hot dry weather the leaves curl upward, causing it to be popularly called the Prayer Plant. West Indian arrowroot is derived from *M. arundinacea*, which has been cultivated for that purpose in tropical America since ancient times.

Like *Calathea*, *Maranta* must have plenty of moisture and protection from strong sunlight and winds. Given well-drained soil and frequent watering, it makes an attractive house plant. Propagation is by root division.

2

1 *Maranta* 2 *Maranta arundinacea* **'Variegata'**

Schefflera

Botanical Family: *Araliaceae*

Ranging in size from fairly low to several metres high (*see* p. 39, The Umbrella Tree) and displaying equally varied leaflets, both in colouration and in shape, *Schefflera* has long been popular as a house plant because of its decorative foliage and tolerance of indoor conditions. It is also useful in gardens, to provide tall background plants or mixed with flowering shrubs. In the wild, some species grow as epiphytic plants attached to the trunks or branches of trees. The leaves are palmately arranged on long stalks in most species.

In time, *Schefflera* may lose its lower leaves and become scandent, in which case pruning will produce new branches and improve the shape. Cuttings often fail to root, and propagation is more easily done by air-layering or pinning longer stems to the ground.

3 & 4

3 *Schefflera farinosa*   4 *Schefflera*

1 & 2

Dieffenbachia
Dumb Cane

Botanical Family: *Araceae*

Nearly all the most commonly seen forms of this widely used native of the Caribbean have been derived from *D. maculata* 'Rudolph Roehrs', a natural mutant discovered in the Roehrs Nursery in New Jersey, U.S.A., in 1936. Now countless varieties are available with leaves that range from light to dark green with white, cream, or yellow markings. As a general rule, those displaying a predominance of white or yellow need more sun for the colour to develop most successfully. The popular name comes from the fact that *Dieffenbachia* contains calcium oxalate crystals which, if swallowed, cause several mouth and throat irritations and may even lead to death.

The plant is easily propagated with stem sections of any length, placed vertically or horizontally in the potting mixture and kept moist.

1 *Dieffenbachia* **'Tropic Snow'**

2 *Dieffenbachia* **'Tropic Snow'** x *Dieffenbachia maculata* **'Rudolph Roehrs'**

Aglaonema
Chinese Evergreen

Botanical Family: *Araceae*

Closely related to *Dieffenbachia*, *Aglaonema* is a genus of twenty-odd species, most of them relatively low plants with attractive variegated leaves. *A. commutatum* is perhaps the most widely used in gardens, often as a ground cover in areas with poor soil and little sunlight, though *A. pictum* is more showy with its glossy, strongly coloured leaves, which often have prominent silver markings. There are countless cultivars, several with strikingly patterned leaves and white stems. They are also popular as house plants thanks to their ability to remain healthy looking for long periods without direct sunlight or good ventilation.

Leafless stem cuttings root easily in well-drained potting soil or when planted directly in a prepared bed. Numerous attractive hybrids of *A. pictum* have been produced by Thai growers in the past few decades.

3

3 **Aglaonema crispum** 4 **Aglaonema pictum cultivar**

4

Sanchezia speciosa (Sanchezia nobilis)

Botanical Family: *Acanthaceae*

Though it has quite attractive stalks of bright yellow, tubular flowers, this South American native is more often grown for its ornamental leaves, which are large, elliptical, and dark green with prominent yellow veins. It grows into a fairly large shrub, up to 2 metres high, though it is often kept lower by frequent pruning, and thrives both in the lowlands and at higher elevations. Since it tends to sprawl, it is usually planted as part of a mixed bed or hedge rather than as an individual specimen. It will grow in shade but needs partial sun to achieve the best leaf colour.

After each flowering, the stalks should be pruned back to encourage denser growth. Cuttings root easily in a light, well-drained potting mixture.

1

2

1 *Sanchezia speciosa*
2 *Sanchezia speciosa*
3 *Graptophyllum pictum*
4 *Polyscias scutellaria* **'Balfourii'**
5 *Polyscias balfouriana* cultivar

3

Graptophyllum pictum

Caricature Plant

Botanical Family: *Acanthaceae*

This small shrub, which is believed to have originated in New Guinea, has patterned ovate leaves that may be green and yellow or bronze, orange, pink, or purplish with paler markings. The colours are not as bright as those of *Codiaeum* or *Acalypha* but can be effective when the plant is used as a low hedge. The flowers appear in short erect panicles and are relatively inconspicuous on most varieties.

Graptophyllum grows best in filtered sunlight or light shade and is easily propagated by cuttings. It should be pruned occasionally when still small to encourage a bushier shape. In temperate climates it requires a warm greenhouse.

4

Polyscias
Panax

Botanical Family: *Araliaceae*

This is one of the most useful foliage plants in tropical gardens, lending itself readily to hedges, massed beds, and background planting, and it also serves as an attractive potted specimen for courtyards and terraces. The leaves may be varying shades of green or variegated and come in a remarkable number of shapes. On *P. scutellaria* 'Balfourii', for example, commonly called the Dinner Plate Aralia, they are large and round; on *P. fruticosa* they are feathery and fernlike and on *P. filicifolia* they are pinnate with deeply lobed margins. Several cultivars can be trained into interesting shapes and grown as instant bonsai.

Most varieties are hardy and will tolerate considerable neglect but require pruning to prevent them from becoming leggy. Propagation is by air-layering (marcotting) or from woody cuttings.

5

1 & 2

Asplenium nidus

Bird's Nest Fern

Botanical Family: *Aspleniaceae*

Native to tropical Asia, this is by nature an epiphyte and can be seen growing wild in the branches of tall trees. With very good drainage, it is also possible to use *Asplenium* in the ground, where it imparts a very lush atmosphere to a jungle-type garden. The long, leathery leaves – narrow on some varieties, very wide on others – grow in a huge rosette and are renewed periodically all year long. In time, a plant can reach a diameter of 1.5 metres.

Asplenium will adjust to full sun but prefers light shade; similarly, in a garden as in nature, it tolerates a dry spell but does best with frequent watering. Propagation is by division or by spores, which often sprout in moist areas around the parent plant.

Xanthosoma

Botanical Family: *Araceae*

Bearing edible tubers rich in starch, *Xanthosoma* has long been noted as a food plant in its native South America. The indigenous people were cultivating several species at the time Columbus made his voyage, and the plant later spread to other tropical regions. At least two species, are also used effectively as ornamentals either in gardens or as potted specimens: *X. lindenii* has long, arrowhead-shaped leaves with boldly contrasting white veins, and *X. violaceum* has dark green leaves with purple veins and purple stems.

In a garden, the plants prefer rich, moist soil and filtered sunlight. The underground tubers spread rapidly, and the plant can be easily propagated by separating shoots and transplanting them to a new location.

3

Platycerium ▼
Stag's Horn Fern
Botanical Family: *Polypodiaceae*

This is a large epiphyte that grows on tree trunks, branches, and rocks and certainly adds appeal to a lush garden landscape. It is not a true ground cover plant but its long antler-like fronds can be grown to drape over rocks and trees, covering a large area. Huge, sterile, kidney-shaped fronds wrap around its roots, trapping leaves and other nutrients, while long, fertile, antler-like fronds hang down, sometimes for a considerable distance. On *P. coronarium*, one of the commonest species, the lower fronds may hang for 1 metre or more; on *P. grande* they are smaller and more complex.

Platycerium prefers filtered shade and a strong host plant where the hanging fronds can be appreciated. It will tolerate a dry spell but does best with regular watering.

Propagation is by means of spores that form on the leaves.

1 **Asplenium nidus** Bird's Nest Fern	3 **Asparagus densiflorus**
	4 **Platycerium**
2 **Xanthosoma lindenii**	Stag's Horn Fern

Asparagus densiflorus ▲
Aparagus Fern
Botanical Family: *Liliaceae*

This is another of those plants like *Selaginella* that looks like a fern but actually isn't. It has tuberous roots and bright green stems with small, modified branches that shoot from the centre of the plant. It is often grown as a pot plant, and the stems are often used in flower arrangements. On the cultivar 'Sprengeri', favoured by florists, the leaves are larger, while on another, 'Myersii', sometimes called the Foxtail Fern, the leaves are short and tightly arranged around the stems so that they look like bushy green tails.

The Asparagus Fern will grow in light shade and prefers moist, well-drained soil, though it will tolerate a period of drought. It is propagated by means of root division.

4

1 & 2

Codiaeum variegatum

Croton

Botanical Family: *Euphorbiaceae*

The first recorded finding of *Codiaeum* was by H.A. Van Rheede, a Dutch official assigned to the Malabar coast of India; a picture of the plant was published in a book on Indian flora in 1686, calling it "tajere-maram", or "similar to the variegation of a snake". The great botanist Carl von Linne classified it as *Croton variegatus* in 1763, though an earlier naturalist, G.E. Rumphius, who had worked most of his life on the Indonesian island of Amboina, had already identified it as belonging to a distinct genus he called *Codiaeum*. The more eminent Linne's classification has persisted in popular use, though the actual *Croton* genus is quite different and is not cultivated as a garden plant.

According to Dr B. Frank Brown in a pioneering work on the subject, it seems likely that *Codiaeum* originated in the Molluca Islands of Indonesia, at first in a form with all-green leaves. This wild variety, *C.* 'Mollucanum', in time produced a "sport" – that is, an offshoot differing from the parent plant – on which the leaves had bright yellow specks and which became known as *C.* 'Aureo Mollucanum', popularly called "Gold Dust". All subsequent varieties, in Dr Brown's opinion, have derived from a single species, their wide range of colours and leaf forms being due to sports and the fact that *Codiaeum* seeds nearly always produce plants unlike their parents.

1 *Codiaeum* 'Ovafolium'   2 **Mixed** *Codiaeum*

The first *Codiaeum* reached England from the East Indies in 1804. Others soon followed, with leaves of different colours and shapes, and the shrub became a common greenhouse ornamental. Many varieties were introduced by the nursery firm of James Veitch and Sons, while others were crossed and originated by growers in Belgium and France. *Codiaeum* crossed the Atlantic towards the end of the 19th century and became the most popular of all potted house plants. The vogue passed, however, largely as a result of the introduction of other ornamentals like *Philodendron* and *Dieffenbachia,* which survived better in heated homes; but it has remained a favourite with tropical gardeners and new hybrids continue to be introduced, particularly in Thailand where they are regarded as bearers of good luck as well as being decorative.

Codiaeum varies in size from about 1 to around 2.5 metres high and also displays at least six distinct leaf shapes, from large and oval to long and narrow. The colour combinations are almost as wide-ranging as the names that have been given to the numerous hybrids.

In a tropical garden, the shrub needs well-drained soil, good sun, and occasional applications of liquid manure. It can also be grown under warm or slightly cool greenhouse conditions in a sunny location. As a house plant, it does not tolerate prolonged exposure to air-conditioning. Propagation is by air-layering or half-woody cuttings; young cuttings will also root in water over a period of time.

3 *Codiaeum* 'Yellow Johanna Coppinger'
4 *Codiaeum* 'Chief'
5 *Codiaeum* 'Stoplight'
6 *Codiaeum* 'Banana'
7 *Codiaeum* 'Super Petra'
8 *Codiaeum* 'Stewartii'

3

4

5

6

7

8

Cyathea
Tree Fern
Botanical Family: *Cyatheaceae*

Like the Stag's Horn Fern, this very beautiful
plant actually belongs in another section –
perhaps Foliage Plants – but is being grouped
with ferns for convenience. It grows quite tall
– 10 to 15 metres in some varieties – on a
thick trunk, with the huge, lacy green fronds
emerging from the top. It does not grow
well in lowland tropical gardens such as
those of Singapore and Bangkok but can be
spectacular at higher elevations or in places
where temperatures do not rise above the
low 80s F. Many of the commonly grown
species, like *C. cooperi*, are natives of
Australia.

Tree ferns prefer moist but well-drained
soil and light shade. They are propagated by
means of spores.

Cyathea Tree Fern

125

4 Vines and Creepers

Clambering up a tall forest tree, spilling dramatically down a hillside, growing over a garden trellis, sometimes clipped and trained into formal shapes, vines and creepers include some of the most decorative of tropical plants.

Among them are such long-established favourites as the multicoloured *Bougainvillea* and the golden *Allamanda,* both of which escaped their original Brazilian habitat several centuries ago and have become a familiar part of both tropical and sub-tropical landscapes everywhere. Others, however, are perhaps less well-known to most gardeners. There is the Jade Vine, for instance, a native of the Philippine jungle, with extraordinary hanging clusters of blue-green flowers, or the equally spectacular New Guinea Creeper which presents a similar chandelier-like display, but in this case of dazzling red-orange blossoms. Newcomers to the tropics are apt to mistake the mauve-blue flowers of *Petrea volubilis* for some form of Wisteria, at least from a distance or until they handle the rough leaves that inspired its popular name of Sandpaper Vine. Also a memorable sight in full bloom are the pungently scented Garlic Vine, the red-and-white Rangoon Creeper, the elegant climbing Gloriosa Lily, and the delicate Blue Pea, whose flowers are used to colour food in Southeast Asia.

Left **Pyrostegia venusta** Flame Vine **growing over a trellis**

Above **Quisqualis indica** Rangoon Creeper

127

1

Bougainvillea

Botanical Family: *Nyctaginaceae*

Evoking not only the true tropics but also sunny coasts as far north as the Mediterranean, *Bougainvillea* was named after the French navigator, Louis de Bougainville, who came across it during an 18th-century visit to Brazil. The showy vine quickly became a garden favourite and has since been extensively hybridized with forms and colours very different from those of the original plant.

By nature, *Bougainvillea* is a climber or scandent shrub with stems that can reach several metres in length, usually clinging with the aid of curved spines. It can also, however, be clipped to form a hedge or trained into tree-like and topiary shapes with sizeable trunks. Most varieties have pale green ovate leaves in pairs; there is also a form with variegated, green-and-white leaves. The true flowers are small, white, tubular, and insignificant, the bright colour being provided by surrounding bracts that appear in profusion. The most common colours are purple or magenta, but cultivated forms are available in all hues, from pure white to orange, pink, and crimson. There are also varieties on which two colours – pink and white, for instance – appear on the same plant, and others with large double bracts. The bracts usually fall of their own accord, but those on the double-bract form remain after turning brown and must be removed by hand to keep the plant from looking unsightly.

B. glabra and *B. spectabilis* are the species that have been most extensively cultivated, and *B. peruviana* is also widely used in dry regions. Crossbreeding and hybridization have produced a wide range of colours, bract sizes, and blooming frequency, along with an equally large number of names, which may vary from region to region.

2

Bougainvillea always needs full sun to flower best and prefers dry or at least very well-drained soil, whether grown in a garden or as a pot plant. Many varieties flower profusely only during a prolonged dry period or, if grown in pots, when water is withheld or limited; they do particularly well in seaside gardens. Hard pruning after flowering promotes bushy growth and more flowers. Most forms are easily propagated by woody cuttings.

1 *Bougainvillea spectabilis* '**Mary Palmer**'
2 *Bougainvillea spectabilis*
3 *Bougainvillea glabra*

3

Ipomoea

Botanical Family: *Convolvulaceae*

This large genus, which includes some 400 species, is best known for its climbers, many of which go under the popular name of Morning Glory. Mostly native to tropical America, they have been introduced throughout the tropical world and in several places have become wild. One of the most popular, often grown over fences as a screen, is *I. indica*, which has deep blue or mauve flowers. Others include the pale mauve *I. pulchella* and the wine-red *I. horsfalliae*. The typical Morning Glory blooms when the sun first strikes it, but *I. alba* (sometimes listed as *Calonyction aculeatum*), the Moon Flower, displays its huge white fragrant flowers only after dark.

Most climbing species are short-lived and must be regrown from seed after a year or so.

Bauhinia

Botanical Family: *Leguminosae*

Bauhinia is best-known in the form of trees and shrubs, but other members of the genus are climbers, some of which add splashes of dramatic colour to gardens large enough to accommodate them. One of the most beautiful is *B. kockiana*, a native of the Malaysian jungle, which produces a frequent display of orange and red-orange flowers, and another, similarly coloured, is *B. bidentata*. Both require a sizeable tree or a stout support on which to grow.

As their forest origins suggest, these climbing species like rich, well-drained soil and shaded roots. They can be propagated by means of woody cuttings or seeds.

1 **Ipomoea indica**   2 **Bauhinia kockiana**

Passiflora

Passion Flower

Botanical Family: *Passifloraceae*

There are some 400 species in this genus, nearly all native to tropical America. They climb quickly by means of tendrils along the stems and have very striking flowers that are complex in shape, sometimes scented, and often beautifully coloured. On *P. caerulea* the flowers are predominantly blue; they are purple on *P. edulis* and red on *P. coccinea* and *P. vitifolia*. The popular name derives from the supposed resemblance of the flower to the crown of thorns worn by Christ. Several varieties are commonly grown in greenhouses, trained on a trellis or wire-netting, but with a support can also be used as an ornamental in gardens.

Most *Passiflora* species require a sunny location and very well-drained soil. Propagation is by cuttings from mature wood.

3 ***Passiflora vitifolia*** Passion Flower

3

Solandra grandiflora (*S. nitida*)

Cup of Gold, Chalice Vine

Botanical Family: *Solanaceae*

This woody creeper can be a spectacular sight when covered with its large, cup-like yellow-gold flowers, which are fragrant in the evening. Flowers appear several times a year, usually in profusion. The leaves are elliptical, slightly pointed at the end, and glossy on the upper surface. The plant grows rapidly and needs a strong support that will show off the blooms to best effect. Another species often seen in gardens is *S. longiflora*, a scandent shrub, with more oval leaves and creamy white flowers. They belong to the same family as the Deadly Nightshade and the Tomato.

Solandra likes well-drained soil and prefers full sun, though it will also grow in partial shade. Propagation is by means of young-growth cuttings.

1 *Solandra grandiflora* Cup of Gold

1

Odontadenia macrantha

Botanical Family: *Apocynaceae*

This woody climbing shrub from South America has slender stems, oblong leaves and fragrant tubular flowers. It makes a handsome climber with regular displays of flowers, reaching a height of up to 18 m.

It needs good sunlight and very well-drained soil. Propagation is by means of cuttings,

2

Quisqualis indica

Rangoon Creeper

Botanical Family: *Combretaceae*

A native of Southeast Asia, this fast-growing climber makes a decorative covering for a trellis or fence. It has woody, vining stems and soft, light green leaves. Periodically – almost continuously in some locations – it has drooping clusters of fragrant flowers with petals that are white when they first open and subsequently turn pink or crimson, all three colours appearing on a single cluster. There is a predominantly pink form and another with double blossoms.

So vigorous is the Rangoon Creeper that it may need occasional hard pruning to keep it within the desired limits. It needs full sun and moist but well-drained soil. Propagation is by means of woody cuttings or root suckers, which are produced freely.

2 ***Odontadenia macrantha***   3 ***Quisqualis indica***
                                    Rangoon Creeper

3

133

Tristellateia australasiae
Galphimia Vine

Botanical Family: *Malpighiaceae*

The bright yellow, star-shaped flowers that appear in sprays almost all the time on this climber from Malaysia and Australia look very much like those of the shrub *Galphimia glauca*, hence its popular name. Each flower consists of five yellow petals and a cluster of short, red stamens; the leaves are pale green and tend to fold along the middle. Given good soil, *Tristellateia* grows quickly and makes a decorative covering for a garden fence or trellis.

It prefers a sunny location and regular manuring. Propagation is by means of cuttings or seeds, which are produced freely.

2

1 *Tristellateia australasiae*   2 *Tristellateia australasiae*

3

Gloriosa superba ▼
Climbing Lily, Glory Lily
Botanical Family: *Liliaceae*

A native of tropical Africa, this is a very ornamental little vine which climbs by means of tendrils at the ends of its long, blade-shaped leaves. The lily-like flowers consist of six narrow red-and-orange petals with crinkled edges, curving upward in a cluster, and six prominent stamens projecting below. The plant grows from underground tubers and may die back from time to time, but new stems soon appear. *G. superba* is the most commonly grown species; *Gloriosa* 'Rothschildiana' has wider petals that are completely red.

    Gloriosa needs rich, well-drained soil and, though it likes to reach up into the sun, prefers to have the shade of nearby plants over its roots.

Clerodendrum

Botanical Family: *Verbenaceae*

Clerodendrum splendens, a vigorous climber native to tropical Africa, is useful in gardens as it likes to grow in the shade, putting forth clusters of bright red flowers. Another climbing species, *Clerodendrum thomsoniae*, the Bleeding Heart vine, is familiar to glasshouse gardeners since it is fairly small and blooms almost continuously as a potted specimen. It has dark green ovate leaves and striking flowers that appear in sprays and consist of large white calyces that set off bright red petals. Flowers are sometimes followed by fruits that are green at first and then turn black. Both these species prefer somewhat dry conditions and at least partial shade. Propagation is by means of cuttings or seeds.

4

5

3 **Clerodendrum splendens**
4 **Clerodendrum thomsoniae**
  Bleeding Heart vine

5 **Gloriosa superba**
  Glory Lily

Pyrostegia venusta ▽
(P. ignea)

Flame Vine, Orange Trumpet Vine

Botanical Family: *Bignoniaceae*

This fast-growing vine from Brazil, which can densely cover a tree or trellis, is a remarkable sight when it bursts into full bloom with a profusion of brilliant orange tubular flowers. It has shiny leaves composed of two or three leaflets and climbs by means of thread-like tendrils. It flowers best in cooler areas and at slightly higher altitudes; in Thailand, for example, it is a common feature of northern gardens but rarely seen in low-lying Bangkok. Another popular member of the same botanical family is *Bignonia magnifica*, a climber that can also be pruned into a bush, which has large tubular purple or mauve-pink flowers.

Both these vines need full sun and very well-drained soil.

2

Allamanda ▲

Golden Trumpet

Botanical Family: *Apocynaceae*

There are about twelve species of *Allamanda*, some woody climbers and others more shrub-like in habit. With their glossy green leaves and almost continuous display of bright yellow, funnel-shaped flowers, they are among the most popular of tropical ornamentals, especially in sandy seaside gardens where they do particularly well. *A. cathartica* is a vigorous climber with large flowers, both single and double; *A. nertifolia* is more of a shrub, and *A. violacea* has pale mauve instead of the customary yellow flowers. There is also a form with silvery-grey leaves. All parts of the plant, including its milky sap, are toxic.

Allamanda is easily propagated by means of woody cuttings.

1

1 **Pyrostegia venusta**
Flame Vine

2 **Allamanda cathartica**
Golden Trumpet

Holmskioldia sanguinea
Cup and Saucer Plant

Botanical Family: *Verbenaceae*

This is actually a scandent shrub, originally from the sub-tropical Himalaya region, which is best grown over some sort of support to accommodate its long, sprawling branches. The unusual flowers consist of a trumpet-shaped corolla which rises from the centre of a large red-orange saucer-shaped calyx (another of its popular names is the Chinese Hat Plant). It grows most vigorously and flowers most brightly during the dry season at slightly higher elevations. There is also a rarer pure yellow form.

Holmskioldia prefers full sun or light shade and well-drained soil. Propagation is by means of woody cuttings, which root easily.

3 *Holmskioldia sanguinea*

3

Congea tomentosa

Botanical Family: *Verbenaceae*

Native to Thailand and Burma, this climber can be an impressive sight in full bloom when growing up a tree or over a strong support in a garden. It has ovate, slightly hairy leaves and seasonal displays of velvety flower bracts that appear in sprays and are either dusty pink or silvery white. It blooms best in regions where there is a pronounced dry season and should be pruned back hard after each flowering. It can also be pruned into a large bush, though this will inhibit blooming since the flowers appear at the ends of the stems.

Congea likes full sun and well-drained soil, though it prefers to have its roots in the shade, especially when young. Woody cuttings root easily.

1 **Congea tomentosa**   2 **Congea tomentosa**

2

1

3

Argyreia nervosa ▲

Elephant Creeper,
Silver Morning Glory

Botanical Family: *Convolvulaceae*

Elephant Creeper is an apt popular name for this native of India, which grows so rapidly that it can cover a sizeable area in a remarkably short time. Given a strong enough support, however, it makes an attractive garden plant with its large leaves that are dark green on top and silvery white below. The rose-coloured flowers are also quite large, 5 to 8 cm. long, but not as conspicuous as that might suggest, since they are covered with white hairs on the outside and tend to get lost among the leaves.

Argyreia needs full sun and well-drained soil. Propagation is by woody cuttings.

Petrea volubilis ▽ ▷

Sandpaper Vine

Botanical Family: *Verbenaceae*

Visually (though not of course botanically) the closest thing to a Wisteria that can be found in the tropics is this woody climber from Central and South America, which periodically puts forth masses of beautiful mauve-blue flowers. Each bloom actually consists of a small dark mauve flower surrounded by larger and paler calyces; the flowers fall quickly but the calyces remain for a week or more.

P. rugosa is a shrub variety which has shorter flower sprays and grows to about 2 metres; *P. volubilis* 'Albiflora' is white-flowering. The popular name derives from the rough, sandpapery texture of the dark green leaves.

5

Petrea likes full sun and the shrub form should be pruned back after flowering. Propagation is by air-layering or woody cuttings, which are slow to root.

4 *Petrea volubilis*
Sandpaper Vine

5 *Petrea volubilis*
Sandpaper Vine

3 *Argyreia nervosa* Elephant Creeper

4

Ficus pumila

Creeping Fig

Botanical Family: *Moraceae*

Ficus is usually thought of as a tree or a shrub. This species, however, is a woody climber which attaches itself by short roots to walls or any other available support and makes a very attractive cover. The most commonly grown form has dark green ovate leaves that are small at first but become quite large and rather coarse-looking when the vine reaches the top of a wall and develops woody stems; at this point, pale green inedible fruits also appear. Fruiting shoots should be removed to encourage more even coverage by the smaller leaves below. There is also a form with variegated leaves known as 'Snowflake'.

The Creeping Fig prefers good sun, especially when it is getting started. Propagation is by means of cuttings, slow to root but fast when they become established.

1 *Ficus pumila* Creeping Fig

1

2

Mansoa alliacea (*Pseudocalymma alliaceum*)

Garlic Vine

Botanical Family: *Bignoniaceae*

This slender climber from tropical America has glossy green leaves that grow in opposite pairs and frequent clusters of pale mauve and white flowers that have a distinct, garlic aroma, giving rise to the popular name. (Other species of the *Bignoniaceae* family also have this disadvantage or, depending on one's tastes, distinction.) For this reason, some gardeners are unwilling to grow it too near open windows and terraces, though in fact the smell is not that strong and it makes an attractive covering for low fences.

It needs full sun and well-drained soil to bloom profusely. Propagation is by means of woody cuttings.

2 **Mansoa alliacea**
Garlic Vine

3 **Mansoa alliacea**
Garlic Vine

3

1

3

Jasminum ▲
Jasmine

Botanical Family: *Oleaceae*

Jasmines include both shrubs and climbers, some of which can be grown even in cool regions. Perhaps the most popular of the tropical climbers is *Jasminum sambac*, an evergreen found covering fences in countless gardens. Sometimes called Arabian Jasmine, it has elliptical, dark green leaves and very fragrant clusters of single or double white flowers. The blooms are used throughout Asia as religious offerings, in floral decorations, and to flavour tea and food. *J. rex*, or Royal Jasmine, is native to Thailand and although not fragrant produces very large, star-shaped flowers that are much more conspicuous than those of *J. sambac*.

Both species like full or partial sun and well-drained soil. They can be propagated by means of woody cuttings or air-layering.

Thunbergia grandiflora
Blue Trumpet Vine

Botanical Family: *Acanthaceae*

The genus *Thunbergia* contains around 100 species, from shrubs to extremely ornamental vines. *T. grandiflora* is perhaps the best known of the latter, a very fast-growing climber with large, slightly rough heart-shaped leaves and long, hanging clusters of either mauve-blue or white trumpet-shaped flowers. *T. laurifolia* has similar but smaller blooms and leaves that are narrow, smooth, and leathery. *T. mysorensis*, which does best in a cooler climate like that of Hawaii or northern Malaysia, has hanging clusters of decorative yellow and brownish-red flowers.

Thunbergia likes full sun or light shade and moist, well-drained soil. Propagation is by cuttings or air-layering.

2

1 **Thunbergia grandiflora**
Blue Trumpet Vine

2 **Thunbergia grandiflora**
Blue Trumpet Vine

3 **Jasminum rex**
Royal Jasmine

4

Antigonon leptopus
Honolulu Creeper, Coral Vine

Botanical Family: *Polygonaceae*

This slender climber, which originated in Mexico but is now found throughout the tropics, has acquired a large number of popular names in addition to the two mentioned above. They include Mexican Creeper, Bride's Tears, Queen's Jewels, and, in Mexico itself, *cadena de amor*, "Chain of Love". It has medium-sized green heart-shaped leaves with wavy edges, climbs rapidly with the aid of tendrils, and produces an almost constant display of floral racemes that may be pure white but are more often some shade of pink. With a proper support, it can grow as high as 10 metres but may need hard pruning to prevent the bottom from becoming bare.

Antigonon needs full sun and humid conditions to flower profusely but can be grown even in quite cool regions. Propagation is by means of woody cuttings.

4 *Antigonon leptopus*
Honolulu Creeper

5 *Antigonon leptopus*
Honolulu Creeper

5

1

Mucuna bennettii
New Guinea Creeper

Botanical Family: *Leguminosae*

Though a different genus, this rampant climber belongs to the same botanical family as the Jade Vine and has similar three-parted leaves and pea-type flowers that hang in clusters from the woody stems. Instead of being blue-green, however, they are a vivid red-orange and thus even more prominent. The vine was introduced to cultivation only in 1940, when seeds collected in the New Guinea jungle were successfully germinated at the Singapore Botanic Garden. It has since become a popular specimen plant, grown over pergolas or other supports where the flowers can be enjoyed from below.

Like the Jade Vine, the New Guinea Creeper prefers its roots to be in the shade and a moist, well-drained soil.

Strophanthus gratus (*Roupellia grata*)

Botanical Family: *Apocynaceae*

This is a scandent shrub from tropical Africa which with regular pruning can be made into a bush, but which also produces long stems that can be trained over a pergola or other support. It has attractive, leathery, olive-green leaves, purplish stems, and frequent clusters of large, waxy, trumpet-shaped flowers that first appear as decorative purple-red buds which are pink when they open. The flowers are long-lasting and can be cut for floral decorations.

Strophanthus needs full sun to flower profusely. Propagation is by means of woody cuttings.

2

1 *Mucuna bennettii*   2 *Strophanthus gratus*
New Guinea Creeper

3

Strongylodon macrobotrys
Jade Vine

Botanical Family: *Leguminosae*

Perhaps no other tropical gardening experience can compare with coming across this native of the Philippine jungle in full bloom. A rampant climber, it grows high into trees or densely covers a trellis with its three-parted leaves, which are purplish or pale green when they first appear and later harden into darker green. The flowers emerge from the woody stems and hang down in large clusters that may be ½ metre long, each one pea-type in appearance and an extraordinary blue-green in colour. They remain beautiful for several days even after they fall and in Hawaii are used to make leis.

The Jade Vine likes moist but well-drained soil and, especially when young, its roots should be shaded. In a garden planting, care should be taken to provide room for the hanging flower clusters to be seen to their best advantage.

3 *Strongylodon macrobotrys* Jade Vine

4 *Strongylodon macrobotrys* Jade Vine

4

Hoya ▼

Wax Flower, Porcelain Flower

Botanical Family: *Asclepiadaceae*

An evergreen vine that likes to attach itself
to trees by means of roots that appear from
the stems, *Hoya* has thick green oval leaves
and is prized by many tropical plant
collectors because of its small, waxy, star-
shaped flowers, which are fragrant at night.
Numerous hybrids have been made from
H. carnosa, some with flowers that are
creamy white with a pink centre and others
on which the flowers are dark pink or red.
On other species, the flowers may be
reddish-brown or maroon, or the leaves
may be intriguingly twisted and crinkled.

    Hoya likes shady, humid conditions and
prefers a host plant on which to climb,
though it can also be trained up wires
and grown as a potted plant.

2

3

1

4

◀ *Philodendron* ▶

Botanical Family: *Araceae*

Philodendron is a large, constantly growing genus, with some 350 species thus far described. While a few are in the category known as "self-heading", which means that the leaves are produced at the top of a stout trunk, most can be found clambering high into the jungles of their native tropical America. To mention only a few of the numerous species, *P. scandens* has heart-shaped glossy leaves that are bronzy when young and later dark green, *P. domesticum* has long dark green leaves, *P. erubescens* has purple stems and leaves that are dark green on top and purple below, and *P. radiatum* has leaves that are plain and heart-shaped when young but become deeply incised as they mature.

All like shady, moist, forest-type garden conditions and can be grown either up trees or as a dense ground cover.

1 *Hoya carnosa* '**Compacta**'

2 *Philodendron* *panduriforma*

3 *Philodendron scandens*

4 *Philodendron radiatum*

5 *Philodendron* hybrid '**Emerald Duke**'

5

Epipremnum pinnatum 'Aureum'

Pothos, Variegated Philodendron

Botanical Family: *Araceae*

This Asian climber, well-known as a house plant and also in tropical gardens, may be found listed under a variety of names. It was called a *Philodendron* for many years, then placed in the genus *Scindapsus,* finally in its own genus. While by habit a vine, clambering up large trees, it makes a very good ground cover for shady areas, where it helps to create a lush, jungle atmosphere and also inhibits weed growth. The oval leaves may be patterned in dark green and yellow or all yellow; when young or growing on the ground, they are fairly small but increase dramatically in size once the plant begins to climb.

    Epipremnum likes filtered sunlight but will adapt to almost total shade. It is easily propagated by means of cuttings.

1 *Epipremnum pinnatum* '**Aureum**'

Monstera deliciosa
Swiss Cheese Plant

Botanical Family: *Araceae*

Monstera was also once lumped among the *Philodendron* species but is now recognized as a separate genus. *M. deliciosa* is the largest of the group, first collected in the wild from Central America and now a popular house plant as well as a striking addition to any tropical garden. At first the climber produces simple, heart-shaped leaves, but as it grows these become increasingly large, dark green, and perforated. On a mature specimen in the tropics, the leaves can be nearly 1 metre long and almost as wide. Periodically it produces a large, creamy white inflorescence consisting of a waxy spathe and spadix; the latter eventually becomes a multi-berried fruit which when ripe has a taste resembling that of pineapple.

Monstera prefers shady, moist conditions and a strong tree or other support against which to grow. Propagation is by means of terminal cuttings.

2

4

3

2 **Monstera obliqua**
3 **Monstera deliciosa**

4 **Variegated** **Monstera**

5 Exotics

To the average gardener in temperate regions, almost any of the specimens in a book about tropical plants is likely to seem exotic. Some, however, tend to stand out in terms of dazzling colour or sometimes bizarre form, to epitomize the popular view of the tropics as a place of botanical surprise and wonder and, simply for that reason, to demand a separate grouping of their own.

Few plants illustrate this better than *Heliconia*, with spectacular flower bracts that may be as glaringly brilliant as a neon sign or as subtly tinted as a watercolour. *Heliconia* has long been used in garden landscapes but new varieties and species are constantly appearing to delight enthusiasts, either developed by growers or discovered in their native tropical American jungles by intrepid plant hunters. *Heliconia* belongs to a larger taxonomic group called the Zingiberales, which also includes a number of other decidedly unusual plants – *Strelitzia*, for example, or Bird of Paradise; bananas, not only the familiar fruiting forms but many with extraordinary flowers or patterned leaves; and members of the ginger family with equally beautiful flowers that may emerge prominently from the ends of leaf stalks or shoot straight from the ground. Then there is *Anthurium*, a member of the Arum family, with richly coloured "flowers" that cover the spectrum from palest pink to delicate mauve; the delicate Blue Ginger, which in fact is not a ginger at all; and the Bat Plant (*Tacca*), which has almost black, cuplike flowers adorned with long black filaments.

Left   ***Heliconia wagneriana***
Above ***Heliconia caribeae***

1

Heliconia

Botanical Family: *Heliconiaceae*

Heliconia, the only genus in the family *Heliconiaceae*, is among the fastest-expanding groups of ornamental plants. Fred Berry and W. John Kress, who published the first book on the subject in 1991, estimated that there were from 200 to 250 named species and almost that many forms or cultivars; in the relatively few years since then, numerous new forms have either been discovered by plant hunters or developed in nurseries to supply an apparently endless taste for novelty among collectors. The Heliconia Society International, founded in 1985, now has members all over the tropical world, and through cut-flowers and potted specimens the plant is becoming familiar to a much wider public.

Heliconia ranges in size from the small *H. stricta* form known as 'Dwarf Jamaican'; through medium-sized, like *H. psittacorum*; to giants like *H. caribeae*, which can reach up to 3 metres. Growing from underground rhizomes, all have erect shoots and leaves that may be vertically arranged like those of a banana, horizontally like those of a ginger, or obliquely like members of the genus *Canna*. On many the most memorable feature, at least as far as gardeners are concerned, is the inflorescence, which may be erect or pendent and consists of relatively insignificant true flowers and brilliantly coloured bracts. Some, however, are grown mainly for their decorative foliage.

The most common colours are bright red or yellow, often a combination of the two, but there are numerous others, ranging from pale pink to very dark wine. To make just a random selection to show the range, one might cite *H. caribeae* 'Black Magic', on which the huge bracts are a dark burgundy colour; *H. caribeae* 'Gold', a similar variety on which they are bright gold;

2

3

4

5

1 *Heliconia collinsiana*
2 *Heliconia indica* **'Spectabilis'**
3 *Heliconia stricta* **'Firebird'**
4 *Heliconia orthotricha* **'Total Eclipse'**
5 *Heliconia* **'Golden Torch'** (*H. psittacorum* x *H. spathocircinata*)

153

H. bihai 'Arawak', mostly red with areas of yellow, pale orange, and green; *H. stricta* 'Dorado Gold', yellow with a slash of pink; *H. orthotricha* 'Edge of Night', rich crimson edged with emerald green; *H. psittacorum* 'Suriname Sassy', a blend of pink, yellow, orange and green; and *H. chartacca* 'Sexy Pink', with long hanging bracts that are pink and pale greenish-blue.

Heliconia is mostly native to the American tropics, many of the most beautiful specimens growing below 1,200 feet but others found at higher altitudes up to 6,000 feet. They are pollinated exclusively by hummingbirds and though natural hybrids are relatively rare, according to Berry and Kress they do occur, as in the popular yellow-gold *Heliconia* 'Golden Torch', a natural cross between *H. psittacorum* and *H. spathocircinata*. Others are appearing in gardens and nurseries where assorted varieties are grown together.

Heliconia needs rich soil, often manured, plenty of water (but good drainage), and sunlight. Each plant stalk blooms but once and should be cut back to the ground afterwards. The larger varieties need plenty of room to expand and so are best suited to larger spaces. Propagation is by means of rhizomes, which spread rapidly.

6 **Heliconia rostrata**

7 **Heliconia psittacorum**

8 **Heliconia caribeae**

9 **H. psittacorum 'Suriname Sassy'**

6

7

8

9

Musa

Banana

Botanical Family: *Musaceae*

Edible bananas have been cultivated throughout the tropics for thousands of years, and countless cultivars have been developed to improve the taste, size, and abundance of the fruit. Other members of the family, however, are grown primarily for their ornamental value, which may derive either from handsome leaves or prominent flowers. *M. acuminata* 'Sumatrana', for example, has foliage that is maroon underneath and spotted with maroon on top; *M. uranoscopus* (*M. coccinea*) has bright red flowers that emerge from the top of the stalk and on *M. ornata* the flowers range from pale lavender to purple. *M. velutina* has decorative pink flower bracts and small, bright pink velvety fruits.

Bananas, flowering or otherwise, need full sun, rich moist soil, and protection from strong winds that may tear the leaves. Propagation is by root division.

10 ***Musa uranoscopus***   12 ***Musa uranoscopus***
11 ***Musa ornata***

1

2

Alpinia

Botanical Family: *Zingiberaceae*

Resembling *Zingiber*, the true ginger, in appearance, *Alpinia* grows in clusters that consist of long stems with often large leaves and, on many, very conspicuous flowers – actually bracts – that appear at the tops of stems. Because of its lush appearance (the stems may reach 2 metres or more in height) and its long-lasting flowers, it is among the most widely used of tropical garden ornamentals. *A. purpurata*, the Red Ginger, has bright red bracts that cover the small white true flowers; among its numerous cultivars are pink and white-flowering varieties, on some of which the flowers are exceptionally large and rounded. On *A. zerumbet*, the Shell Ginger, the bracts are pink and shell-like with a red and gold interior; a smaller variety called *A. zerumbet* 'Variegata' has striking green-and-yellow foliage.

Alpinia likes full sun or partial shade, moist, well-drained soil, and protection from strong winds. Propagation is by means of root division.

1 *Alpinia purpurata*   2 *A. zerumbet* 'Variegata'
Variegated Ginger

Etlingera elatior
Torch Ginger

Botanical Family: *Zingiberaceae*

This native of tropical Asia has gone through a somewhat bewildering number of botanical names; in some reference works it is listed as *Phaeomeria magnifica*, in others as *P. speciosa*, in still others as *Nicolaia elatior*. The above seems to be the accepted designation at present. The plant puts forth ginger-like stems that can reach a height of 3 metres from underground rhizomes; the flowers also emerge from the ground and consist of ornate red or pink waxy bracts arranged around many small flowers in a way that does resemble a torch. *E. hemisphaerica* has shorter stalks with purple leaves and tulip-shaped inflorescences.

The Torch Ginger likes full sun or light shade and rich, well-drained soil; it should be grown in a site where the flowers can be seen at the bottom of the stalks. Propagation is by clump division.

1 *Etlingera elatior*
Torch Ginger

2 *Etlingera elatior*
Torch Ginger

3

Curcuma ▲

Turmeric

Botanical Family: *Zingiberaceae*

Native to the Indo-Malaysian region, *Curcuma* has long been domestically cultivated because of the spice turmeric, which is extracted from its roots and widely used in the preparation of curries. The root is also a source of yellow dyes and traditional medicines. Although they die back seasonally, particularly in regions with a long dry season, they are becoming increasingly popular as a tropical garden plant due to the striking flowers that appear on several species. On *C. longa*, the commonest, these consist of large green and pink bracts, whereas on *C. roscoeana*, popularly known as the Jewel of Burma, the bracts are gold at first and turn orange with age. There are also other varieties with smaller white or white and chocolate coloured bracts.

Curcuma likes a shady, jungle-like environment, with moist, well-drained soil. Propagation is by root division.

3 *Curcuma roscoeana*
Jewel of Burma

Dichorisandra thrysiflora ▼

Blue Ginger

Botanical Family: *Commelinaceae*

Despite its popular name, this belongs not to the ginger but to the spiderwort family. A succulent, herbaceous plant native to tropical America, it has cane-like stalks around 1.5 metres in height that grow from underground stems, spirally arranged leaves that are dark green with silver markings, and large terminal clusters of flowers that are dark blue-mauve, a rare colour in tropical gardens.

Dichorisandra does best in slightly cooler regions, where temperatures do not rise above the low 80s F.; thus, while quite common in Hawaii, it is seldom seen in Bangkok, Singapore, or Bali. It prefers filtered light and moist, well-drained soil. Propagation is by tip cuttings or root division.

4 *Dichorisandra thrysiflora* Blue Ginger

4

159

1

Calathea ▲

Botanical Family: *Marantaceae*

Calathea is usually grown for its highly ornamental leaves and appears in the chapter on Foliage Plants. At least two species, however, are prized by gardeners because of their exotic flowers that appear prominently among large green leaves. *C. burle-marxii* 'Blue Ice', named for the celebrated Brazilian garden designer and plant collector, has lavender-tipped white flowers surrounded by bracts that are an extraordinary crystalline blue in colour. *C. insignis*, also a native of tropical America, is much taller, up to 1½ metres, and produces small flowers within long, flat yellow bracts that resemble a rattle (the popular name is the Rattlesnake Plant).

Both species are tender plants, preferring shade, moist, well-drained soil, and protection from strong wind. Propagation is by root division.

Zingiber ▼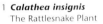

Ginger

Botanical Family: *Zingiberaceae*

The best-known member of this genus, which includes about 80 species, is *Z. officinale*, the commercial source of the ginger root so widely used in food and medicine. Also popular in tropical gardens are several species with colourful and interestingly shaped flowers that, like those of the Torch Ginger, emerge straight from the ground among the leaf stalks. On *Z. spectabile*, the Beehive Ginger, these are tight clusters of waxy yellow bracts surrounding small yellow and white flowers, and on *Z. zerumbet* they are even more compact, green when they first appear and later bright red. *Z. zerumbet* 'Darceyi' is a smaller cultivar with variegated leaves.

Zingiber is adaptable to a variety of growing conditions, either sun or shade, but prefers a protected situation and moist, well-drained soil. Propagation is by root division.

1 **Calathea insignis**
The Rattlesnake Plant

2 **Flower heads of *Zingiber* spectabile** Beehive ginger

2

Tapeinochilus ananassae
Indonesian Wax Ginger

Botanical Family: *Costaceae*

This remarkable plant from Indonesia and New Guinea has tall, branching, ginger-like stems that rise from the ground and large leaves. The inflorescences, which also come out of the ground at the base of the clump, consist of small yellow flowers surrounded by tightly arranged, bright red bracts which have thorns on their tips. These pineapple-shaped structures are very prominent and make a dramatic splash of colour in a shady part of a garden.

A jungle plant, *Tapeinochilus* prefers half-shade and rich, moist soil; it does not grow well in very hot regions with a pronounced dry season. Propagation is by root division or by cuttings taken just below a joint on the stem.

3 *Tapeinochilus ananassae* Indonesian Wax Ginger

3

Anthurium

Botanical Family: *Araceae*

The genus *Anthurium*, the largest in the Aroid family, contains more than 700 species. Some are grown for their decorative foliage, but perhaps the best known to plant lovers who live outside the tropics are those with beautifully coloured "flowers", which are actually large spathes, or leaf bracts, surrounding the fleshy flower spikes. The best known is *A. andraeanum*, sometimes called the Flamingo Flower, which has bright red glossy spathes. Discovered in 1876 in Colombia, it proved a sensation in European glasshouses and nurseries and was used to produce numerous hybrids with spathes in varying shades of pink and pure white. On *A. scherzerianum* the spathes are smaller but equally dramatic, in shades of orange or red. A species discovered as recently as 1972, in Panama, has mauve-coloured spathes and is now being widely cultivated.

Anthurium requires filtered sunlight and a moist but very well-drained growing medium; in pots, this is often a mixture of soil, fiber, and broken bricks, with regular applications of liquid fertilizer. Propagation is by means of offshoots from the parent plant.

1 **Anthurium andraeanum**

2 **Anthurium andraenum hybrid**

3 **Anthurium lilacinum**

1

3

2

4

5

6

Costus
Spiral Ginger

Botanical Family: *Costaceae*

This group includes a number of ginger-like herbaceous plants that rise up to a metre or so from the ground and have spiralling leaves. *C. speciosus*, the Crape Ginger, is one of the most popular with tropical gardeners; it has dark green, spear-shaped leaves grouped around the top of the stalks and large white flowers with a red-orange throat that emerge from terminal pinkish-red cones. *C. speciosus* 'Variegatus' has very beautiful white-striped foliage. On *C. curvibracteus* the flower bracts are orange; on *C. spicatus* they form bright red cones. The flowers on *C. malortieanus* are less prominent but the large, furry leaves are subtly patterned.

Costus prefers light shade and moist but well-drained soil; some species may die back during a prolonged dry season but will return with the rains. Propagation is by clump division.

4 **Costus speciosus**
Crape Ginger

5 **Costus curvibracteus**

6 **Costus speciosus**
'Variegatus'

Strelitzia
Bird of Paradise

Botanical Family: *Strelitziaceae*

Named in honour of Charlotte of Mecklenburg-Strelitz, German-born queen of Great Britain's George III, the striking plants in this small family are native to South Africa but are now widely grown in places like Hawaii and Southern California. The best-known is *S. reginae*, around 1 metre tall with long, leathery, banana-like leaves and extraordinary long-lasting flowers that are orange and blue. *S. nicolai* is much larger, growing several metres high, with dense clusters of palm-like leaves and large flowers that are white and blue.

Strelitzia likes full sun, very well-drained soil, and prefers dry rather than very moist climates. Propagation is by seeds (which must be soaked) or root division.

Tacca chantrieri
Bat Plant, Devil Flower

Botanical Family: *Taccaceae*

The somewhat sinister names of this native of Southeast Asia are due to its curious flowers. These rise from clumps of glossy green leaves, are almost black and are surrounded by large purplish-black bracts with long, thin appendages. *T. integrifolia*, is similar in appearance. *T. leontopetaloides*, the Polynesian arrowroot, is grown commercially for its starchy tubers.

Tacca likes filtered sunlight, some protection from heavy rain, and rich, moist, well-drained soil. Propagation is by root division.

1 **Strelitzia reginae**
  Bird of Paradise

2 **Tacca chantrieri** Bat Plant

3 **Garden with Strelitzia**

6 *Ground Covers and Other Plants for Similar Use*

Ground covers and other plants for similar use in massed, mostly low-growing arrangements are as essential to tropical gardeners as they are to those elsewhere. They serve to link and set off the larger landscape elements and to hold the soil during the heavy rains that are characteristic of most tropical regions for long periods. They also bring interesting colours and textures into a garden.

Creeping plants like *Alternanthera*, *Lantana*, or the hardy *Wedelia*, for instance, can be used not only to cover large areas quickly but also to achieve patterns of contrasting colour for visual interest. Roberto Burle-Marx, the noted Brazilian designer, was a pioneer in this field, working like a painter to create vivid sunny landscapes almost entirely composed of low plants with coloured foliage. Slightly taller plants with distinctive leaves like a dwarf variegated *Pandanus*, spiky *Tradescantia*, purple and silver *Hemigraphis*, and jewel-like Coleus also add to the variety of such compositions. As an alternative to the fast growing, shade-loving vines and creepers, like *Epipremnum* or *Philodendron*, one or more of the numerous ferns can be used to fill shadier beds, with handsomely patterned *Peperomia*, or the glossy green leaves and white flower spikes of *Spathiphyllum*, a popular house specimen in temperate climates but a vigorous bedding plant in the tropics.

Left   **A courtyard with *Epipremnum aureum* and *Tradescantia spathacea* used as ground covers**
Above   ***Selaginella***

1

Alternanthera ficoidea ▲
Joseph's Coat

Botanical Family: *Amaranthaceae*

Growing only around 20 cm. high, spreading rapidly in a suitable location, and available in a variety of colours, this native of South America is a very useful ground cover and can be seen used as such in numerous tropical gardens. The white flowers are insignificant, but the small oval leaves may be bright or dark red, pink, green, yellow, or variegated, sometimes a combination.

Alternanthera requires well-drained soil and good sunlight for the colours to develop properly; it will grow in the shade but often remains green. It can be clipped to maintain the desired height but this should be done regularly. Propagation is by means of cuttings or division of larger plants.

Solenostemon scutellarioides
Coleus ▼

Botanical Family: *Labiatae*

This has developed into a highly varied and immensely decorative group of low-growing herbs or shrubs (which in some reference works now appears under the botanical name of *Plectranthus*). They have brightly coloured leaves in a wide range of patterns and spikes of small purple-blue flowers. The most common colours are red, bronze, or yellow, but many beautifully patterned hybrids have been produced. The leaves of some all-green varieties native to Southeast Asia are used in flavouring food.

Coleus grows most vigorously in regions where temperatures do not rise above the mid-80s F. and likes full or filtered sunlight and well-drained soil. It is easily propagated by means of cuttings.

1 *Alternanthera ficoidea* 'Variegata'

2–6 **A sample of the range of decorative Coleus leaf patterns**

2

3

5

4

6

169

1

Orthosiphon stamineus (*O. aristatus*)

Cat's Whiskers

Botanical Family: *Labiatae*

This low-growing (about 75 cm. high) herb is native to Southeast Asia. It has attractive, dark green leaves and frequent tall sprays of flowers that are distinctive because of their exceptionally long stamens – which give rise to the popular name. The usual flower colour is white, but there is also a form with pale mauve blooms. The plant is used as a treatment for diabetes in traditional Chinese medicine. With a little pruning to encourage bushy growth, especially after flowering, it makes a very good ground cover.

Orthosiphon likes full or filtered sunlight and well-drained soil. Propagation is by means of cuttings.

2

Ophiopogon

Botanical Family: *Liliaceae*

Two species of *Ophiopogon* are useful when a grass-like cover with an interesting texture is desired. *O. japonicus*, sometimes called Mondo Grass, is a solid green cover that grows in tufts and likes shady areas. *O. intermedius*, or Lillturf, does best in filtered sunlight, has broader leaves with prominent white margins and occasionally produces sprays of small white flowers. Neither variety spreads rapidly and they require little maintenance beyond regular watering.

Propagation is by division of plants.

1 *Orthosiphon stamineus*   2 *Ophiopogon*
Cat's Whiskers

3

Peperomia

Botanical Family: *Piperaceae*

Members of the same family as the vine grown commercially for black pepper, these low-growing, fleshy plants are often grown as potted specimens because of their often very ornamental foliage. With proper care, some varieties can also be used as ground covers, especially in areas where temperatures do not rise above the low 80s F. *P. obtusifolia*, sometimes called the Baby Rubber Plant, has thick green and cream or yellow leaves, *P. argyreia* has heart-shaped leaves that are striped like a watermelon and *P. caperata* has quilted, dark red leaves on red stems.

    Peperomia usually requires shade and moist but very well-drained soil; overwatering or flooding will cause the roots to rot. Propagation is by plant division or stem cuttings.

4

3 ***Peperomia obtusifolia***   4 ***Peperomia argyreia***
'Variegata'

◀ *Liriope muscari*

Botanical Family: *Liliaceae*

This is a clustering plant that spreads like grass and has long leathery leaves which may be solid green or green-and-white striped. From time to time, it produces spikes of white or purple flowers, but these are not very prominent. Native to southern China, it grows equally well in tropical and sub-tropical gardens and makes an effective cover.

Liriope likes good drainage and regular watering; a good deal of sun is required for the variegated cultivars to show their best colour contrasts. Propagation is by means of plant division.

| 1 *Liriope muscari* | 3 *Episcia reptans* |
| 2 *Episcia cupreata* | |

2

3

Episcia ▲

Botanical Family: *Gesneriaceae*

This creeping plant, native to South America, is often seen grown in hanging pots where its attractive foliage and small but sometimes bright flowers can be admired at close quarters. Of the two most commonly grown varieties, *E. cupreata* has soft, ovate leaves that are reddish or bronze with silver markings, whereas on *E. reptans* the leaves are green with lighter markings. There are also cultivars with reddish-brown or rose-pink foliage. The flowers are usually crimson. The plant puts out long runners, at the end of which a new plantlet develops, covering the ground fairly rapidly.

Episcia likes light shade, well-drained soil, and protection from heavy rains.

1

2

Zebrina pendula ▼

Wandering Jew

Botanical Family: *Commelinaceae*

This creeping plant from Mexico is a popular
house plant in temperate countries and is
also frequently used as a ground
cover in the tropics. It has fleshy, ovate-
oblong leaves that are dark green with
silvery-white stripes on top and purple
below, and it spreads rapidly by rooting
stems. There are several forms, among
them one called 'Quadricolor' on
which the leaves are more colourfully
striped.

    Zebrina needs some sun for the colour
to develop best, though too much sun can
make the leaves look faded. Either in the
ground or in a hanging pot, it prefers
moist but well-drained soil. Propagation
is by means of cuttings.

◀ Wedelia trilobata

Botanical Family: *Compositae*

Few ground covers are used more widely
in tropical gardens than this hardy native
of the Caribbean, which grows in sun or
shade, tolerates most soil conditions as
well as salty winds, and spreads very rapidly
over a large area. It has shiny, dark green
leaves and bright yellow daisy-like flowers,
and spreads by means of creeping, rooted
stems. Left alone, it can become rather
straggly but it responds well to frequent
pruning and can even be clipped into
attractive forms; it is also effective when
planted so that it trails down a wall from
a raised container.

    Other than pruning, *Wedelia* requires
little in the way of maintenance. It is
easily propagated by terminal cuttings,
which root quickly.

Lantana montevidensis ▲ (L. sellowiana)

Botanical Family: *Verbenaceae*

Though *Lantana* is usually found as a shrub,
sometimes an invasive one, this species is
trailing and a very useful cover for sunny
areas. It has rough, dark green, ovate leaves
and an almost continuous display of rosy-
mauve flowers. There are other forms on
which the flowers are yellow, white, pink,
or a combination of colours. It spreads by
means of creeping stems which can reach
4 metres in length and can thus be effective
in a raised container or hanging basket as
well as in the ground.

    Lantana needs full sun and well-drained,
slightly dry soil to do well; it will die back
in a place that is frequently flooded or water-
logged. Propagation is by means of cuttings
or rooted stems.

3

1 **Wedelia trilobata** (on the high wall)	3 **Zebrina pendula**
2 **Lantana montevidensis**	

Pandanus pygmaeus

Botanical Family: *Pandanaceae*

Pandanus can be very large, almost the size of a small tree. This species, however, is a dwarf, growing to only about 50 cm., with very attractive long narrow leaves brilliantly striped with bright yellow. It makes a dense, colourful cover in the right location, requiring little maintenance, though the edges of the leaves have very sharp prickles which should be taken into account when planting it along pathways. Another low-growing species, *P. amaryllifolius*, has smooth, dark green leaves, often used in Thai cooking.

    P. pygmaeus needs full sunlight for the colour to develop properly; in a shady spot the leaves will become all green. Propagation is by means of plant division.

Portulaca

Botanical Family: *Portulacaceae*

This decorative, ever-blooming succulent from South America prefers much drier conditions than many gardens in the true tropics can provide, but it is still often seen in rockeries, raised containers, and hanging baskets where good drainage can be ensured. It has thick stems and fleshy leaves. One form produces yellow, pink, red, and purple flowers on the same plant, while another has large, double magenta blooms. The flowers open in the morning when sun strikes the plant.

    As suggested above, *Portulaca* requires full sun and excellent drainage. Propagation is by means of cuttings and seeds.

1 **Pandanus pygmaeus**   2 **Portulaca grandiflora**

Cuphea

False Heather

Botanical Family: *Lythraceae*

This low-growing native of South America, which reminds some of heather, is found in both tropical and sub-tropical gardens. *C. hyssopifolia*, often used as a bedding plant along pathways or to cover slopes, grows to about 60 cm. and has tiny, dark green leaves and frequent displays of flowers that may be mauve, white, or pink. *C. ignea* has bright red flowers.

Cuphea prefers full sun but is tolerant of most soils; it grows better in slightly cooler climates than in places like Singapore and Bangkok, where it can become straggly and bare around the bottom. Propagation is by means of seeds or cuttings.

3 **Cuphea hyssopifolia**

3

177

Phyllanthus myrtifolius

Botanical Family: *Euphorbiaceae*

This is a very attractive low-growing shrub, particularly useful in raised beds or along the edges of an artificial pond. Its long, slender branches have small, fern-like leaves and often sprout roots when they touch the ground. The branches droop as the plant grows, making it an effective cover when planted among larger specimens. As it grows taller, the lower parts may become bare, in which case hard pruning will produce a bushier effect.

Phyllanthus will tolerate most conditions but prefers full sun and slightly dry soil, making it a goodchoice for rock gardens. It is propagated by means of cuttings or, more rapidly, removal of rooted branches.

2

Pedilanthus ▲

Botanical Family: *Euphorbiaceae*

This is a succulent shrub from tropical America which grows to about 50 cm. in height on stems along which the leaves are arranged on each side. The most commonly grown species is *P. tithymaloides*, with green leaves and occasional small sprays of red flowers; there are various forms, one with variegated leaves and another that remains under 30 cm. and has very tightly arranged leaves. Some gardeners object to its spiky, rather rigid appearance but it can be useful in a massed planting and tolerates neglect.

Pedilanthus needs full sun and well-drained soil; pruning will encourage more branches and promote a bushier appearance. Cuttings root easily.

1

1 *Phyllanthus myrtifolius*   2 *Pedilanthus tithymaloides*

Sansevieria trifasciata
Mother-in-law's Tongue

Botanical Family: *Agavaceae*

Tolerating neglect and even hard treatment, this native of South Africa has long been a popular house plant. In gardens, it is often relegated to obscure and difficult places that receive little attention, though if properly cared for it can make an attractive cover. It has leathery leaves that grow from thick rhizomes and, on some species, occasional sprays of fragrant flowers. *S. trifasciata*, the most commonly grown, occurs in numerous cultivars; one, for instance, has sword-shaped leaves about 50 cm. high, dark green with bright yellow borders; on another the leaves are silver-coloured. *S. cylindrica* has rounded leaves and is grown more for its curiosity value than its beauty.

Sansevieria will grow in sun or partial shade but need dry conditions. Propagation is by division of the rhizomes.

3 *Sansevieria trifasciata* Mother-in-law's Tongue

3

Tradescantia spathacea
(*Rhoeo spathacea*)
(*R. discolor*)

Boat Lily, Moses in a Boat

Botanical Family: *Commelinaceae*

Originating in Central America, this attractive plant is now used everywhere in the tropics as a ground cover or to edge pathways . It has a short stem and a spiky rosette of leaves that are reddish-purple below and greenish-purple on top. White flowers appear between boat-shaped bracts in the axils of the leaves, giving rise to the popular names. It usually grows to about 35-40 cm.; but a dwarf form is about half that size. There is a variegated form with yellow stripes. The entire plant is poisonous.

Given well-drained soil and full or lightly filtered sunlight *Rhoeo* needs little care and spreads quickly. It can easily be propagated from seedlings, or by plant division.

2

1

Hemigraphis alternata (H. colorata)

Botanical Family: *Acanthaceae*

This is a very low-growing cover from Indonesia which can also be planted attractively in hanging baskets. It has spreading, rooting stems, small white flowers, and ovate leaves which are purplish with a metallic sheen on top and darker purple underneath. On a cultivar known as 'Exotica' the leaves are puckered and turn down at the edges, while another species, *H. repanda*, has long, thin, strongly notched leaves which are similarly coloured.

    Hemigraphis prefers full or filtered sunlight and good drainage; given these conditions, it will spread rapidly and require little maintenance. Propagation is by terminal cuttings or rooted stems.

3

Tradescantia pallida (Setcreasea pallida)

Purple Heart

Botanical Family: *Commelinaceae*

Also listed as *Setcreasea purpurea*, this Mexican native is a trailing plant with fleshy stems and slightly hairy lance-shaped leaves, both a vivid shade of purple. Ocassionally, it has small flowers that are paler purple. The branches break easily, so it needs some protection, but well-sited it is colourful as a ground cover or in a rock garden.

    Setcreasea likes well-drained but not completely dry soil, and full sun is necessary for the most intense colour. Pinching back encourages bushier plants and a thicker cover. Propagation is by means of cuttings.

1 **Tradescantia spathacea**	3 **Hemigraphis alternata**
2 **Tradescantia spathacea** Boat Lily	4 **Tradescantia pallida** Purple Heart

4

Selaginella

Botanical Family: *Selaginellaceae*

Selaginella is not a true fern, although it branches in a way that gives it a frond-like appearance. It grows wild in many parts of the tropical and sub-tropical world – some thirty species are native to Malaysia – and can make a very decorative cover in shady, moist parts of a garden. One of the most beautiful is *S. willdenowii*, popularly called the Peacock Fern, on which the delicate leaves appear blue when lit from a certain angle. Another common variety is *S. ciliaris*, which grows like a clubmoss.

    Selaginella needs shade and plenty of water; it tends to die back during prolonged dry spells. Propagation is by division of older plants or rooted stems.

Nephrolepis

Botanical Family: *Nephrolepidaceae*

Members of this hardy, easy-to-grow genus have been standard house plants since Victorian times and are equally common as tropical garden ground covers. The best known species is *N. exalta*, especially the cultivar 'Bostoniensis', or Boston Fern; it has long, green pinnate leaves and grows to about 50 cm.; another, known as the Lace Fern, is more delicate and is best grown in pots or hanging baskets. Also popular is *N. biserrata*, which has very long fronds, more than a metre in length.

    Nephrolepis does best in filtered sunlight in moist but well-drained soil; generally, it needs less water than most other ferns. It spreads by runners and propagation is by means of division.

1 **Selaginella**   2 **Nephrolepis exalta**

Adiantum
Maidenhair Fern

Botanical Family: *Adiantaceae*

Among the most elegant of all ferns, and very popular as house plants in temperate countries, or in rockeries or well-prepared beds in tropical gardens, Maidenhairs have square or triangular leaflets on black wiry stems and range from diminutive to very large. Some are native to Southeast Asia, but most come from tropical America and seem to prefer slightly cooler places. *A. tenerum,* is very graceful and is found in several forms, one with frilly leaflets. Of the giant Maidenhairs, two of the best known are *A. peruvianum* and *A. trapeziforme.*

All prefer shade or very light sun and must have perfect drainage; they do better when protected from heavy rains. Propagation is by division of rootstocks.

3 **Adiantum tenerum**
Maidenhair Fern

4 **A. tenerum at the Botanic Gardens, Singapore**

◀ *Syngonium*

Arrowhead Vine

Botanical Family: *Araceae*

Native to the New World tropics, the genus *Syngonium* contains around 30 species, of which the most commonly grown is *S. podophyllum*. The leaves change to a marked degree as the plant grows; when young they have only two or three lobes and are often variegated, but with age they have multiple lobes and become entirely green. Numerous cultivars have been developed, some with foliage that is cream-coloured and tipped with dark green at the edges. Like *Epipremnum* and *Philodendron*, this is a rapid climber but can be used as decorative ground cover and makes an excellent indoor plant. Propagation is by means of cuttings.

Spathiphyllum ▶

Botanical Family: *Araceae*

The genus *Spathiphyllum* includes around 40 species, most of them from tropical America. With its broad, dark green leaves and prominent, snow-white spathes, it has become a popular house plant, but in a tropical garden it makes a very attractive ground cover for areas that are shady or receive only filtered sunlight. It will, however, adapt to quite strong sun. The most commonly grown species are *C. wallisii*, *C. cannifolium*, and *C. floribundum*, which vary mostly in leaf size.

    Spathiphyllum always requires moist but well-drained soil and, once established in the ground, will spread rapidly; some of the larger hybrids multiply more slowly but need ample room in which to spread their leaves. Propagation is by means of offshoots.

1 *Syngonium podophyllum*   2 *Spathiphyllum wallisii*

1

7 *Water Plants*

Water plays a major role in garden landscapes everywhere. This is particularly true in the tropics, where formal pools, natural-looking lakes and ponds, artificial cascades, even large water jars in a courtyard help to lower temperatures as well as providing focal points of interest. In traditional Balinese temples and palaces, such water features were a major part of the architecture, and they also figured prominently in the larger gardens of Thailand, Burma, and Indo-China. Today, even quite small tropical gardens try to incorporate one somewhere, partly to please the eye and partly to create a place where some of the many decorative tropical water plants can be brought into the landscape.

There are countless varieties of *Nymphaea*, or water lilies, that open by both day and by night. For gardeners fortunate enough to have a really large pond, few specimens can rival in dramatic effect the spectacular *Victoria amazonica* from Brazil, with its enormous, platterlike leaves that can reach 1 metre or more in diameter, but even a relatively small jar is adequate to cultivate the equally exotic Lotus, sacred to Buddhists and Hindus alike. Additional aquatic plants include the Water Hyacinth and the Water Poppy, tall, feathery *Cyperus papyrus*, the slender, reedlike *Typha angustifolia,* and the elegant Water Canna with its tassels of mauve flowers.

Left   **A lotus-filled pond in a Balinese garden**
Above **Lotus grown in a water jar**

Nymphaea
Water Lily

Botanical Family: *Nymphaeaceae*

As almost every gardener knows, water lilies are by no means confined to the tropics but can be found almost everywhere, even surviving winter freezes. All have spreading rootstocks and groups of leaves that generally float on the surface of the water, though in some tropical varieties the leaves are taller; the flowers rising above may be single or double and occur in a wide range of colours, including blue. Nearly all those seen in tropical gardens are hybrids, generally of *N. lotus*, *N. capensis*, or *N. mexicana*, and may be either night-or day-blooming, both large and small; some cultivars also have decoratively patterned leaves.

Water lilies need full or at least a half day's sun to thrive and bloom profusely. They can be planted in the muddy bottom of ponds or in pots that are sunk in a water jar. Propagation is by seeds or by root division.

1 *Nymphaea rubra*

1

2 *Nymphaea*

3 *Nymphaea capensis*

4 *Nymphaea mexicana* **hybrid**

5 *Nymphaea mexicana* **hybrid**

Nelumbo nucifera

Lotus

Botanical Family: *Nelumbonaceae*

1

Once classified in the family *Nymphaeaceae*, the famous Lotus has now been given one of its own, which contains only two species. The most famous is a native of Asia, growing from India to China; the other is *N. lutea*, a yellow-flowering variety, which comes from North America. *N. nucifera* has large pink or white five-petalled flowers and round leaves, covered with a network of fine hairs that causes raindrops to roll off them; both flowers and leaves rise on long stalks a considerable distance above the water. The plant has been adopted as a sacred symbol by both Buddhists and Hindus.

Nelumbo is a vigorous grower and in a garden pond needs a container or restraint to keep it within the desired area; it can be very effective when grown in a large water jar. Propagation is by seed or division of the creeping rootstocks.

2

Victoria amazonica
Giant Water Lily

Botanical Family: *Nymphaeaceae*

Seeds from this celebrated water lily were germinated at Kew in 1846, but it first bloomed three years later at Chatsworth, where the Duke of Devonshire built a special glasshouse for it. The plant was originally called *V. regia*, but later acquired its present name. A native of the Amazon region, it has leaves that are at least 2 metres in diameter, turned up at the edges and covered with sharp spines on the underside; the protruding ribs that give stability to the huge leaf supposedly inspired the glasshouse designs of Joseph Paxton at Chatsworth and Kew. The large, fragrant flowers are white when they open at dusk and reddish-pink when they close in the morning.

The plant needs full sun and a very large pond, so is not suitable for most small or medium-sized gardens; it grows most vigorously during the rainy season.

1 **Nelumbo nucifera** Lotus
2 **Nelumbo nucifera** Lotus

3 **Victoria amazonica**
Giant Water Lily

3

Pistia stratiotes
Water Lettuce

Botanical Family: *Araceae*

This is the only floating Aroid, which looks like no other member of that large and varied family. The pale green leaves grow in a small, lettuce-like rosette, spread out or even flat on the water if the plants are widely spaced but upright if many are grown close together. Flowers are produced, but they are very small and usually hidden. Masses of *Pistia* can be a hazard, as they have been on the upper reaches of the White Nile, but it can be quite decorative when grown in a confined space like a small pool or a water jar.

    Pistia needs full sun and tolerates sub-tropical as well as tropical climates. It propagates itself by means of offsets, which can be removed for a new planting.

1

Trapa bicornis
Water Chestnut

Botanical Family: *Trapaceae*

This low floating plant is grown commercially in several parts of Asia for its horned, dark-brown fruit, which appears below the water surface and, popularly known as Water Chestnut, is used in a number of dishes. When they reach the surface, the diamond-shaped leaves form an attractive rosette and make an interesting, non-invasive ornamental specimen in water gardens, either in ponds or large containers.

    It can be grown from fresh fruits completely submerged in water, though germination is slow. It prefers full sunlight.

1 **Pistia stratiotes**
Water Lettuce

2 **Trapa bicornis**
Water Chestnut

2

Eichhornia crassipes
Water Hyacinth

Botanical Family: *Pontederiaceae*

This native of tropical America has been widely introduced as an ornamental elsewhere, often with unfortunate results. In new habitats, the floating plant quickly escaped gardens and became a noxious weed that clogs waterways because of its remarkably rapid growth. It remains, however, a very attractive plant when grown in a confined space like a water jar. It has thick, fleshy green leaves and buoyant stalks and periodically produces prominent stalks of blue-mauve flowers.

The Water Hyacinth likes full sun and prefers shallow to deep water. It spreads by means of offshoots, which can be used for propagation.

3 *Eichhornia crassipes*
Water Hyacinth

4 *Eichhornia crassipes*
Water Hyacinth

3

4

Thalia geniculata

Water Canna

Botanical Family: *Marantaceae*

A native of tropical America, this is a very decorative specimen to plant in shallow water or in wet soil. It rises on stalks about 1.5 metres tall and has large, pale green, *Canna*-like leaves. The flowers are purple with violet spots and appear prominently at the ends of the stalks in dangling clusters.

Thalia thrives in full sun at lower elevations. Propagation is by means of seeds or root division.

1 **Thalia geniculata**
Water Canna

2 **Thalia geniculata**
Water Canna

2

1

Cyperus

Botanical Family: *Cyperaceae*

Two species of these ornamental, grass-like plants are often used to provide height in or along the edges of garden ponds. *Cyperus alternifolius*, sometimes called the Umbrella Plant, is perhaps the most common. A native of Madagascar, it grows to a little over 1 metre tall on stalks, at the tops of which are clusters of bright green, grass-like leaves; on a variegated form the leaves are striped with white. Much larger is *C. papyrus*, the Egyptian Papyrus from the Nile region, which can grow to 2 metres and has very elegant, feathery, drooping leaves.

Both varieties prefer sun and *C. papyrus* needs either a large pond or some kind of constraint since it spreads rapidly. Propagation is by means of root division.

3 **Cyperus alternifolius (left) and C. papyrus (right)**

3

◄ *Typha angustifolia*

Botanical Family: *Typhaceae*

Sometimes incorrectly called a Bulrush, this is an attractive tall plant that can provide height and a certain sense of formality to a garden pond. It grows on straight stalks up to 2 metres in height, has stiff, narrow leaves at the base of the stems, and dense, brown-coloured flower spikes that are often used in floral arrangements since they last for a long time when dried.

Typha likes full sun and because of its size is probably best for larger ponds. Propagation is by means of root division.

Acrostichum aureum ►

Giant Mangrove Fern

Botanical Family: *Pteridaceae*

This large, fern-like plant grows wild in mangrove swamps and often appears along the edges of garden ponds as well, carried by microscopic spores. Growing up to 3 metres high, it has long, leathery fronds, normally green but often covered with reddish brown spores.

Though essentially a wild plant and eventually becoming too large for the average garden, *Acrostichum* can nevertheless be used effectively in a landscape, particularly one that strives for a natural effect.

1 *Typha angustifolia*   3 *Acrostichum aureum*
2 *Typha angustifolia*

8 Palms and Palm-Like Plants

No other family evokes the tropics more powerfully than the *Palmae* in one of its many forms, whether the graceful Coconut Palm arching over a beach or the stately Royal Palm along an avenue.

Many of the nearly 4,000 palm species grow in sub-tropical or temperate regions and were familiar to anyone who had visited the Mediterranean coast or the Middle East. Even tropical specimens were popular Victorian house plants and were on display at Kew's enormous Palm House, which opened in London in 1848. However, what came as a revelation was the sight of palms growing naturally and in such profusion everywhere. "It takes a little time", wrote an American visitor to 19th-century Java, "for the temperate mind to accept the palm-tree as a common, natural, and inevitable object in every outlook and landscape; to realize that the joyous living thing with restless, perpetually trashing foliage is the same correct, symmetrical, motionless feather-duster on end that one knows in the still life of hothouses and drawing-rooms at home."

Immensely tall or relatively low, feathery or fanlike, massive or dainty, there is a palm, it would seem, to suit almost any garden, and the problem is to make a choice. In addition, there are plants that are palm-like in their graceful arrangements of stems or leaves, together with the bamboos, which can make a similar contribution to garden landscaping.

Left   **Coconut Palms in the garden of a Thai resort**
Above **Fan Palm**

1

Cocos nucifera
Coconut Palm

This is undoubtedly the most famous member of the family *Palmae*. It is also the most useful; when fresh, the nuts are a major source of food, the dried flesh (or copra) yields commercial oils, the fronds are used for thatching, the trunks for timber. Its origin is uncertain, but by human or natural distribution it is cultivated throughout the tropical world, mainly in coastal regions. Wild coconut palms may grow up to 30 metres on trunks crowned with a tuft of wide-spreading pinnate leaves, but most cultivated varieties are much lower, some producing fruit when only 1 metre or so tall.

Coconut Palms are best suited to fairly large gardens that can accommodate their wide leaf span; thought should also be given as to where the nuts will fall if not removed. They prefer full sun but will grow in a wide range of soils and tolerate exposure to strong sea winds. Propagation is by seed.

1 *Cocos nucifera*	2 *Cocos nucifera*
Coconut Palm	Coconut Palm

2

Borassus flabellifer
Palmyra Palm

Throughout Southeast Asia and southern India, this tall, stately palm is almost as common as the coconut. It is grown in huge plantations for the sweet sap obtained from the unopened inflorescence, which is then transformed into an alcoholic beverage known as "toddy". As with the coconut, the leaves are used for thatching and the trunk for timber; the initial taproot produced by the seed is regarded as a delicacy. It is too large for the average garden but can be attractive in parks.

The Palmyra Palm needs full sun and well-drained soil. It is propagated by seeds.

3 *Borassus flabellifer* Palmyra Palm

3

2

Oncosperma tigillarium
Nibung Palm

This is another palm only for sizeable gardens and parks. It grows in clumps of as many as 40 slim trunks up to 20 metres tall, topped with crowns of feather-shaped leaves that droop gracefully on both sides of the leaf stalk. Unfortunately, the trunks are covered with sharp spines, which make them useful as a protective wall but somewhat difficult for a gardener. Another species, aptly named *O. horridum*, is even more thoroughly armed with spines that make penetration of the clump almost impossible.

The Nibung Palm likes full sun and moist, somewhat heavy soil. Propagation is by division or by seed.

1 *Oncosperma tigillarium*
Nibung Palm

2 *Oncosperma tigillarium*
Nibung Palm

Corypha umbraculifera
Talipot Palm

This native of Sri Lanka is a real giant, suitable only for large gardens but quite spectacular when planted in a site where its massive shape can be appreciated. It grows on a single trunk to about 30 metres and has huge, fan-shaped leaves that may be 3 metres wide. The palm only flowers and fruits when it is about 40 years old and then it dies. The inflorescence, however, is memorable, the largest of any flowering plant, rising about 4 metres above the leaves and displaying some 10 million flowers. *C. utan*, native to India and Burma, is similar but somewhat less massive.

Corypha needs full sun; it is tolerant of many soils but prefers moisture. Propagation is by means of seed.

3 *Corypha umbraculifera*
Talipot Palm

3

Areca catechu
Betel Nut Palm

This fast-growing, solitary palm is commercially cultivated throughout Southeast Asia because of its fruit, which has a mildly narcotic effect. Betel-nut chewing has waned in popularity, but the palm is still widely seen, often planted for its decorative qualities. It grows very rapidly and has a tall, slender trunk, ringed with leaf scars and topped by a crown of dark green leaves; the large, bright red fruit appears just below the crown. In a garden, it can be grown as an elegant solitary specimen that does not take up much room or in groups.

Areca prefers light shade and moist but well-drained soil. It is easily propagated by seed.

Veitchia merrillii
Manila Palm, Christmas Palm

This very attractive, single-stemmed palm, a native of the Philippines, is now widely used in tropical gardens and along streets. Its pale, smooth trunk, slightly swollen at the base, grows to a moderate height of about 5 metres; it has feather-shaped, dark green leaves that curve outward and prominent inflorescences just below the crown of leaves. The bright red fruits, which in some regions appear in December, are thought to resemble Christmas decorations and are thus responsible for one of the palm's popular names.

The Manila Palm likes full sun and moist soil. Propagation is by means of seed, which germinate rather slowly.

1 **Areca catechu**
Betel Nut Palm

2 **Veitchia merrillii**
Manila Palm

Dypsis lutescens
(Chrysalidocarpus lutescens)
Golden Cane Palm

Originating in Madagascar, this is an extremely useful tufted palm, widely grown in both pots and gardens. It grows in thick, bushy clumps with slim, ringed trunks that curve slightly outward and graceful feather-shaped leaves on which the petioles are yellow when the palm is planted in full sun. It makes an excellent focal point in a garden and can also be used as a screen, growing to about 2 metres in height.

Sun is necessary for the yellow leaf markings but the palm will also grow in partial shade and is a very tolerant potted plant. Propagation is by means of rooted offshoots.

3 **Dypsis lutescens** Golden Cane Palm

3

Caryota
Fishtail Palm

This genus is distinguished by its two-pinnate leaves which droop gracefully and resemble fishtails. It has a smooth grey trunk, ringed with scars where old leaves have fallen off; offshoots are produced around the main trunk to form a lower cluster. The flowers, which appear only once before the palm dies, are quite decorative, as are the red or black fruits which follow them. *C. mitis*, the species most often seen in gardens, grows to about 4 metres and with its dense foliage makes a good screen. *C. urens*, sometimes called the Wine Palm because its fruit is used to make a beverage, is taller, around 10 metres, and has exceptionally long, hanging inflorescences.

Caryota will grow in sun or semi-shade and does well in almost any kind of soil as long as it is moist. Propagation is by means of seeds or offshoots.

Bismarckia nobilis
Bismarck Palm

This is an exceptionally beautiful solitary palm with very large, fan-shaped leaves that are blue or silver-grey in colour. It grows very slowly on a massive grey trunk that can eventually reach 10 to 15 metres in height. Leaf colours vary considerably, the most prized being a remarkable blue-grey. It comes originally from Madagascar, where wild specimens are now a rarity.

Bismarckia needs full sun and a site where its wide-spreading leaves can be seen to their best advantage. Propagation is by seeds, which germinate in around 2 to 3 months and display their colour on the first leaves.

1 *Caryota mitis*   2 *Bismarckia nobilis*

1

2

Hyophorbe lagenicaulis ▼
Bottle Palm

The reason for the popular name of this unusual palm is immediately apparent in the bulging bottle-like form of its light grey trunk. The stiff, feather-shaped leaves, sometimes twisted and recurved, emerge from the top of the "bottle". The palm originated in the Mascarene Islands but as a result of extensive collecting it is now almost extinct in the wild. Another variety, *H. verschaffeltii*, the Spindle Palm, is narrow at the bottom and bulges in the middle, though not so prominently as the Bottle Palm.

The Bottle Palm needs full sun and a solitary location where its shape can be appreciated. Propagation is by seed.

Copernicia macroglossa ▲
Petticoat Palm

This fan palm, native to the Caribbean and South America, gets its popular name because the upper part of its trunk is always covered by hanging leaves, which remain for many years after they have died. It is a fairly low species, growing to a height of only around 7 metres on a solitary trunk; before they die and form the "petticoat", the leaves are erect, fan-shaped and deeply cut.

The Petticoat Palm needs full sun and well-drained soil. It is propagated by means of seed, though young plants are very slow-growing.

1 *Copernicia macroglossa*   2 *Hyophorbe lagenicaulis*
Petticoat Palm   Bottle Palm

3

Ptychosperma macarthurii ▲
Macarthur Palm

This is one of the most commonly used garden palms, especially in Southeast Asia. A native of Australia, it grows in clumps of smooth, greenish-grey trunks, each with a crown of dark green leaves; periodically it produces long panicles of cream-coloured flowers, which are followed by very decorative bright red fruits. Groups of the palm planted closely together can make an effective screen in a small garden and potted specimens look good on terraces and patios

The Macarthur Palm will grow in either sun or shade but likes a moist, well-drained soil. Propagation is by seeds, which germinate quickly and are often found sprouting around the base of a clump.

Rhapis ▼
Lady Palm

This native of southern China is perhaps the most popular of all palms for indoor use and is equally familiar in gardens, especially in shady areas where other specimens are difficult to grow. It grows in clumps of many slender trunks, each covered with dark fibres, and has fan-shaped leaves divided into a number of leaflets. It can reach about 2 metres in height but is more often lower. The two most common species are *R. excelsa* and *R. humilis*, which are very similar in appearance; there are several ornamental cultivars, including one with variegated leaves.

Rhapis will grow in dense shade as well as sun but prefers moist soil. Propagation is by means of rooted offshoots.

3 *Ptychosperma macarthurii*	4 *Rhapis humilis*
Macarthur Palm	Lady Palm

4

Cycas
Botanical Family: *Cycadaceae*

Large and fern-like with stiff, shiny, green leaves, Cycads are among the oldest plants in the world. They usually consist of a thick trunk that increases in height very slowly over the years, and a crown of pinnate leaves that appear in whorls once or twice a year. There are a number of species, but probably the most common ones used in tropical gardens are *C. rumphii*, a relatively small variety that becomes around 1 metre in diameter; the larger *C. circinalis*, which resembles a small palm; and the still larger *C. revoluta*, the trunk of which may reach 3 metres or more. Cycads are propagated from seeds.

A close relative of the Cycads is *Zamia*, which resembles them in growing habits but on which the leathery leaflets are larger and often oval-shaped.

1 **Cycas**

1

Phoenix

Date Palm

This genus of mostly single-stemmed palms with feather-shaped leaves contains about 17 species. *P. dactylifera*, a native of North Africa grown throughout the Middle East, is the one commonly known as the Date Palm, but a close relative, *P. canariensis*, from the Canary Islands, is also grown for its delicious orange fruit. *P. reclinata*, the African Date Palm, grows in clumps rather than singly and is most attractive since each of the trunks leans slightly outwards. All these are fairly large; more popular in small gardens and also as house plants are *P. roebelenii*, the Pygmy Date Palm, a native of Laos and southern China, and *P. loureirii*, both of which have delicate, fern-like leaves and grow to only about 2 metres in height.

Date Palms like full sun and slightly dry soil. Propagation is by seed.

2 ***Phoenix roebelenii*** Pygmy Date Palm

2

Elaeis guineensis
Oil Palm

Grown commercially in huge plantations, this native of Africa is an important source of vegetable oils, which are extracted from the fruit. About 20 metres tall when mature, it is too large for the average garden but can be a useful addition to more spacious landscapes such as resorts, where its thick trunk and spreading crown of feather-shaped leaves can be seen to advantage. The old leaf bases that cover the trunk provide an ideal habitat for epiphytes like ferns and orchids, giving the palm an added advantage.

The Oil Palm prefers sun and high humidity. Propagation is by seeds, which germinate very slowly.

1 *Elaeis guineensis* Oil Palm

Roystonea
Royal Palm

This is a popular palm for planting along avenues or in large gardens. Native to the Caribbean, is has a straight, smooth, light grey trunk that, when mature, can reach around 25 metres, with a crown of very large, dark green pinnate leaves; on some species the trunk is slightly swollen towards the bottom, especially when young. The two most commonly grown species are *R. regia*, sometimes called the Cuban Royal Palm, and *R. elata*, the Florida Royal Palm.

Roystonea is a true giant and should be used only where its stately form can be appreciated from a distance. It needs full sun and fairly high humidity. Propagation is by seed.

2 *Roystonea* Royal Palm

2

Cyrtostachys renda
Sealing-wax Palm

This is a tufted palm, which means that it produces offshoots and grows as a group of slender trunks, each with its own crown of leaves. What makes this species stand out is its bright red leaf sheaths, which appear down nearly the full length of the trunk as well as along the leaf stalk; the colour is similar to that of old Chinese sealing wax, giving rise to the popular name. The leaves are stiff and feather-shaped, and a fully grown clump is not too large for a small garden.

Cyrtostachys needs full or partial sun and moist soil. Propagation is by means of offshoots.

1 **Cyrtostachys renda**
Sealing-wax Palm

2 **Cyrtostachys renda**
Sealing-wax Palm

3

Lodoicea maldivica
Coco-de-mer, Double Coconut

This is one of the curiosities of the palm family, sure to be a centre of interest in a garden large enough to accommodate it. Native only to the Seychelle Islands in the Indian Ocean, it grows on a solitary trunk to about 25 metres, has very large fan-shaped leaves which droop at the leaf-ends. It produces a fruit which takes five years to ripen and contains the world's largest seed. This looks like a two-lobed coconut and can weigh up to 40 pounds.

The Double Coconut is a rare palm, the seeds expensive and very slow to germinate. Once established, it prefers full sun and high humidity.

3 **Fruits of** *Lodoicea maldivica*

4 *Lodoicea maldivica* Double Coconut

4

Livistona
Fan Palm

The genus *Livistona* contains about 30 species. *Livistona chinensis*, The Chinese Fan Palm, originating in southern China, has attractive fan-shaped leaves with long, drooping tips, a characteristic that has led some to call it the Fountain Palm; it grows on a solitary trunk to a height of about 3 metres. *Livistona australis*, the Australian Fan Palm, has similarly drooping leaf ends but grows considerably taller, up to 30 metres. Particularly dramatic in appearance is *L. rotundifolia*, which has very wide circular "fans".

Livistona likes full sun but will grow in partial shade. Propagation is by means of seed.

Licuala

This a genus consisting of about 100 species, most of them medium-sized tropical fan palms. One of the most popular is *L. grandis*, the Ruffled Fan Palm, which grows to about 1.5 metres on a solitary trunk and has large, fan-like, regularly pleated leaves that have notched edges. Less commonly seen in gardens but more beautiful is *L. orbicularis*, on which the glossy green leaves are less deeply notched and circular. *L. spinosa* and *L. ferruginea* resemble *L. grandis* except that the leaves are divided from the base into leaflets with squared-off ends.

Licuala will grow in sun or partial shade but needs protection against strong winds. For the species mentioned, propagation is by means of seed.

1 *Livistona rotundifolia* 2 *Licuala ferruginea*
Fan Palm

Wodyetia bifurca
Foxtail Palm

This beautiful solitary palm from Australia has a crown of long fronds on which the leaflets radiate from all sides of the leaf stalk, giving them a feathery appearance that has inspired the popular name. The trunk ranges from light grey to brownish grey, is swollen slightly at the middle, and may grow to about 20 metres though it is usually lower. Both male and female flowers are produced on the inflorescence, followed by orange-red fruits.

Preferring full sun, the Foxtail Palm is drought-tolerant but grows faster in moist soil. Propagation is by means of seeds, which take two to three months to germinate.

3

Johannesteijsmannia altifrons

This is an exceptionally beautiful palm from Malaysia. Growing to a height of about 2 metres, it has no trunk but enormous, undivided leaves that rise straight from an underground stem and are used as superior thatching in its native land. At present it is not commonly seen in gardens, but its attractive form and relatively compact shape make it eligible for wider use.

It prefers a shady location and needs protection from strong winds which will damage the large leaves. Propagation is by seed, which must be fresh.

3 *Wodyetia bifurca*
Foxtail Palm

4 *Johannesteijsmannia altifrons*

4

Dypsis decaryi
(Neodypsis decaryi)
Triangle Palm

This native of Madagascar is easily recognizable because of the unique triangular shape formed by the leaf bases at the top of its trunk, which is otherwise round. The very long, gracefully arching leaves are feather-shaped and silvery green in colour, making the palm stand out prominently in a garden. It is of medium size, growing to about 3 metres, and therefore useful both in gardens and as a potted specimen. Because of its growing popularity and the export of seeds, it has become rare in its original habitat.

Dypsis needs full sun but is very tolerant of soil conditions and periods of drought. It is propagated by means of seed.

| 1 **Dypsis decaryi** | 2 **Dypsis decaryi** |
| *Triangle Palm* | Triangle Palm |

Ravenala madagascariensis
'Traveller's Palm'
Botanical Family: *Strelitziaceae*

This famous native of Madagascar is not, of course, a palm at all. It belongs instead to a family related to the banana, specifically to the *Strelitziaceae* which includes the Bird of Paradise (*see* p. 164). The popular name derives from the fact that drinkable water accumulates at the leaf bases to assuage the thirst of anyone in need. A dramatic feature in traditional tropical gardens – though too large for a small one – its enormous, banana-like leaves fan out from a tall trunk, reaching up to 20 metres in height. Green fan-shaped flowers appear from within the leaves, and many offshoots are produced around the main plant, forming a large clump unless they are removed.

The Traveller's Palm will grow in full sun or partial shade and needs some protection against strong winds that may tear the leaves. It is easily propagated by offshoots.

Bamboo

Botanical Family : *Gramineae*

Going under the popular name of Bamboo are a number of genera, all belonging to the huge *Gramineae*, or grass, family. These include not only the genus *Bambusa*, but also *Arundinaria*, *Phyllostachys*, *Schizostachyum*, *Dendrocalamus* and others. Most grow in full or partial shade, on canes that may be short but are more often very tall. The larger varieties are not generally suited to small gardens since they can become invasive, not to mention the continual problem of leaf drop. Given enough space, however, many can be extremely decorative and contribute much to a tropical atmosphere.

To give only a small selection, *Bambusa vulgaris* 'Aureo-variegata' has canes that are striped with green and bright yellow, while on *Phyllostachys sulphurea* they are golden-coloured. *B. ventricosa*, known as Buddha's Belly, is fairly low-growing, with canes that are swollen between the internodes.

1

2

1 ***Phyllostachys sulphurea***
Bamboo

2 ***Phyllostachys sulphurea***
Bamboo

3 ***Bambusa vulgaris***
Bamboo

4 ***Thyrsostachis siamensis***
Bamboo

5 ***Atroviolacea***
Black Bamboo

4

5

3

6

7

Dendrocalamus giganteus, native to Southeast Asia, is enormous, with canes that can grow to 35 metres and may be 35 cm. in diameter. *Fargeria nitida (Arundinaria nitida)*, by contrast, is relatively small and has distinctive black canes. A further attractive bamboo-like plant is *Arundinaria variegata*, the White Bamboo Grass.

8

6 **Arundinaria variegata** White Bamboo Grass	9 **Bambusa ventricosa** Bamboo
7 **Arundinaria variegata** White Bamboo Grass	10 **Bambusa ventricosa** *Bamboo*
8 **Bambusa multiplex** **'Variegata'** Bamboo	11 **Dendrocalamus** Bamboo

9

10

11

223

9 Orchids

Though there is no precise figure for the total number of orchid species, most authorities estimate it as being somewhere between 20,000 and 25,000, making *Orchidaceae* by far the largest family of flowering plants. To this must be added at least 50,000 or so registered hybrids, which are being continually produced at the rate of about 1,000 per year.

Many people think that orchids are limited to the tropics, but this is by no means the case. Species can be found growing in the high Himalayas, in sandy deserts, in every one of the American states. The great majority, however, are indeed native to the warm, humid regions of the world; the island of New Guinea has the greatest number of any single place, followed by Colombia, Brazil, Costa Rica, Borneo, and Java. Borneo alone, according to a recent study, has between 3,000 and 4,000 species, and quite a few others probably remain undiscovered.

Most temperate zone orchids are terrestrial, meaning that they grow in the ground, while tropical species are more likely (though not always) to be epiphytic, growing on trees and occasionally on rocks. The flowers range in colour through the entire spectrum, vary greatly in form, and may be obscure or quite spectacular – on *Grammatophyllum speciosum,* the largest orchid in the world, the inflorescences can be 2 metres long, each bearing fifty or more large gold and dark red flowers.

Left   ***Grammatophyllum speciosum*, the largest orchid in the world**
Above   ***Vanda* hybrid**

1

2

3

4

Orchids began to attract Western collectors in the early 19th century and soon became a passion among those who could afford the necessary facilities for their cultivation. Adventurous hunters were dispatched to the most remote areas to search for new species, sometimes destroying their natural habitat in the process, and occasional public auctions were held for prized specimens. Sometimes a hint of scandal spiced such events. According to Tyler Whittle in *The Plant Hunters,* "When the long-sought-after orchid *Dendrobium schroederi* was sold at Protheroe's in London, the sale rooms were packed because it was a condition of sale that, in order to pacify the native tribe which had first owned the orchid, the human skull in which it grew had to be bought as well."

Later, towards the end of the century, hybrids replaced wild orchids as the main focus of interest, and to some extent this is still true today. The following selection is limited to only a few of the many genera but will, perhaps, give an idea of the family's immense variety and beauty.

5

1 *Arachnis flos-aeris*
Black Scorpion

2 *Ascocenda*
kwageokchoo

3 *Arundina majula rimau*

4 *Dendrobium*

5 *Arundina graminifolia*

Phalaenopsis

Popularly known as Moth Orchids, these are among the most beautiful and graceful of orchids and thus very popular with collectors. The genus comprises about 70 species, nearly all epiphytic and tropical, and many hybrids with exceptionally large and delicate flowers that seem to float on their long stems.

Generally the plant has a short stem and thick, leathery, dark green leaves that may be quite large. The slender flowering stem, which can grow to 80 cm. in length, may have as many as twenty flowers. *P. amabilis*, native to Indonesia and New Guinea and one of the most commonly grown species, has large flowers that are white flushed with yellow and a complex inner structure that is yellow striped with red. Other varieties have pure white, pink, or rose flowers. *Phalaenopsis* species have also been crossed with other genera such as *Vanda*, *Renanthera*, and *Ascocentrum* to produce a wide variety of flower shapes and colours.

Phalaenopsis does best in full shade and must never dry out, though water should not be allowed to collect in the growing tip. It also needs a porous, very well- drained potting mixture and, in places with heavy rainfall, occasional spraying with fungicide. Regular applications of liquid fertilizer are also beneficial.

1

2

1–5 *Phalaenopsis* **hybrids**

3

4

5

229

1

1 *Dendrobium* **hybrid**   3 **Wild** *Dendrobium*
2 *Dendrobium* **hybrid**

2

Dendrobium

One of the largest orchid genera, this consists of more than 1,500 species and far more hybrids. They are epiphytic and generally consist of a group of pseudobulbs, or stems, of varying length, with alternate leaves; the flowers appear on either racemes or in clusters and are often very striking in both wild and hybrid species. They are relatively easy to grow and thus are particularly popular with collectors, either in a controlled glasshouse environment or outside in a tropical garden.

Found growing wild are *D. crumenatum*, the Pigeon Orchid, which has sprays of fragrant white flowers; *D. chrysotoxum*, with bright yellow, waxy flowers; *D. pulchellum*, on which the flowers are cream with maroon spots, and *D. anosmum* (*D. superbum*), with fragrant mauve flowers. Countless *Dendrobium* hybrids have been produced, with long-lasting flowers in almost every colour. At the Singapore Botanic Gardens, a number have been specifically created for the important cut-flower industry and also in honour of such notable visitors as Queen Elizabeth II and Margaret Thatcher.

These orchids are generally grown in pots or baskets – sometimes tied to trees – in a medium that allows for perfect drainage; tightly packed *Osmunda* fern roots and broken bricks are often used for this purpose. They need at least a half-day's sun, preferably morning, and, in the case of hybrids, regular applications of liquid fertilizer. Repotting is recommended for most every two years or so to remove dead roots. Some of the wild species need a prolonged dry spell to bloom well.

3

1 *Vanda tricolor*   2 *Vanda tricolor*

3

Vanda

The genus *Vanda* contains some 70 species and a large number of hybrids, many of which are grown commercially for the cut-flower trade. Mostly epiphytic, they are monopodial in growth, which means they have a central stem with one growing tip, the flower stalks emerging from the stem between the leaves in each row; aerial roots are also produced from the stems, which in some varities can grow quite tall. The flowers are highly varied in appearance and colour and abundantly produced. *Vanda* can be crossed with a number of other orchid genera, among them *Arachnis*, *Ascocentrum* and *Renanthera*, and there are also natural crosses within the genus itself as well as man-made ones.

In the wild there is the the beautiful *V. caerulea*, with rare blue or blue-mauve flowers, and *V. sanderana*, both of which have been used extensively in hybridization. The most famous natural cross, between *V. teres* and *V. hookeriana*, was discovered in her Singapore garden by Miss Agnes Joaquim in 1893; it was subsequently registered under her name and is today Singapore's national flower.

These orchids are generally easy to grow. They like a good deal of sun – at least half a day – a loose potting mixture such as broken brick and charcoal, and regular applications of liquid fertilizer.

3 *Vanda* 'Miss Joaquim'   4 *Vanda caerulea*

4

1

2

3

Cattleya

To the average person, this is probably the best known of all orchids because of its showy, often very large flowers and beautiful colour range. The genus consists of about 65 species, nearly all epiphytic and native to Central and South America, particularly Brazil, as well as thousands of hybrids.

Cattleya can be roughly divided into two types – those which are two-leafed, or bifoliate, and produce relatively small but numerous flowers at the top of each pseudobulb, and those which are one-leafed, or labiate, and produce fewer but much larger flowers. The latter group has been used most extensively for hybridization and crossbreeding and is the one most often cultivated by collectors; more Cattleya hybrids are registered than for any other orchid genus. The flowers generally have a tubular lip and large, spreading sepals and petals; colours range from pure white to deepest purple.

Cultivation of Cattleya is similar to that of Dendrobium, except that it seems to prefer slightly cooler weather. A mixture of firmly packed Osmunda and charcoal, broken bricks or pot shards is often used, with repotting every two years or so as the plant outgrows its containers. Too much shade will reduce flowering, but too little may result in yellowish leaves; during the growth period it needs plenty of water and high humidity.

1 *Cattleya* '**Alma Kee**'   3 *Cattleya* '**Queen Sirikit**'
2 *Cattleya* '**Lucky Strike**'   4 *Cattleya*

4

Paphiopedilum

Commonly known as Slipper Orchids or Lady's Slippers, this genus contains about 30 species, as well as thousands of hybrids that have become well-known house plants and that are often incorrectly called *Cypripedium*, actually a distinct group. The popular names were inspired by the characteristic pouch-like lip that is a prominent feature of the often large and spectacular flowers.

Usually terrestrial, meaning that they grow in the ground, *Paphiopedilum* is native to a region that extends from China through Southeast Asia to New Guinea. Wild species are now regarded as endangered and trade in them is prohibited by a convention signed by over ninety countries. They have leathery leaves that may be attractively mottled, and the flowers appear, often singly but sometimes in clusters, at the top of stalks. To cite but two examples, on *P. barbatum* the dorsal sepal of the flower is white striped with purple and the petals and lip are varying shades of purple; on *P. lowii* yellow-green is added to the colour combination and the petals droop elegantly.

Most of the hybrids do best under slightly cool conditions. They like light shade, a very well-drained potting mixture, and protection from heavy rains.

1

2

1 *Paphiopedilum* 'Shirén'   2 *Paphiopedilum*

Oncidium

This is an extremely large genus with some 750 species, occurring in the New World from South Florida to Argentina. They are mostly epiphytic, though a few grow in the ground and some are found on rocks. Generally they have thick, stem-like pseudobulbs, each with two leaves that vary considerably in size and shape, and sprays of often showy flowers, mostly in shades of yellow and brown though other colours like green, red, or magenta may appear. A characteristic of most *Oncidium* flowers is their prominent, fiddle-shaped lip. On some varieties, the inflorescences can be very long, each bearing a hundred or more flowers.

Many *Oncidium* hybrids have been produced, some popularly known as "Dancing Dolls" with long sprays of bright yellow flowers. *O.* 'Golden Shower', a cross between *O. flexousum* and *O. sphacelatum*, was the first *Oncidium* hybrid to be developed at the Singapore Botanic Gardens, in 1940, and has become a popular producer of cut-flowers.

Considering the wide range of species it is difficult to be precise about cultivation. Generally, however, they like fairly dry growing conditions, with a very well-drained potting mixture, and a bright, well-ventilated situation.

3 **Oncidium sphacelatum**

3

Bibliography

Amranand, Pimsai and Warren, Willam *Gardening in Bangkok* (revised ed.), Bangkok 1994

Bar-Zevi, David *Tropical Gardening*, New York 1996

Berry, Fred and Kress, W. John *Heliconia: An Identification Guide*, Washington, D.C. 1991

Bown, Deni *Aroids: Plants of the Aroid Family*, London 1988

Brown, R. Frank *A Codiaeum Encyclopedia: Crotons of the World* and *The Cordyline*, Valkaria Tropical Garden, Florida, 1992 and 1994

Bruggeman, L. *Tropical Plants and Their Cultivation* , London 1957

Chan, C.L., Lamb, A., Shim P.S., and Wood, J.J. *Orchids of Borneo*, The Sabah Society, Kota Kinabalu, in association with the Royal Botanic Gardens, Kew, 1994

Clay, Horace F. and Hubbard, James C. *The Hawai'i Garden: Tropical Exotics*, Honolulu 1977

Cronin, Leonard *Key Guide to Australian Palms, Ferns and Allies*, Chatswood, New South Wales, 1989

Eggenberger, Richard and Mary Helen *The Handbook on Plumeria Culture*, Houston 1985

Eiseman, Fred and Margaret *Flowers of Bali*, Singapore 1988

Gibbons, Martin *Identifying Palms*, Edison, NJ, 1995

Gilliland, H.B. *Common Malayan Plants*, Kuala Lumpur 1958

Graf, Alfred Byrd *Tropica: Color Cyclopedia of Exotic Plants and Trees*, East Rutherford, N J, 1981

Greensil, T.M. *Gardening in the Tropics*, London 1964

Hawkes, Alex D. *Orchids: Their Botany and Culture*, New York 1961

Holttum, R.E. and Enoch, Ivan *Gardening in the Tropics*, Singapore 1991

Kiaer, Eigil *Indoor Plants*, London 1961

Kuck, Loraine E. and Tongg, Richard C. *Hawaiian Flowers and Flowering Trees*, Tokyo 1960

Lotschert, W. and Beese, G. *Tropical Plants*, London 1983

McMakin, Patrick D. *A Field Guide to the Flowering Plants of Thailand*, Bangkok 1988

Macmillan, H.F. *Tropical Planting and Gardening*, London 1935

Menninger, Edwin A. *Flowering Trees of the World*, New York 1962

Merrill, Elmer D. *Plant Life of the Pacific World*, Tokyo 1981

Miyano, Leland *Hawai'i, A Floral Paradise*, Honolulu 1995

Oliva-Esteva, Francisco, and Steyermark, Julian, A. *Bromeliaceaes of Venezuela*, Caracas 1987

Polunin, Ivan *Plants and Flowers of Singapore*, Singapore 1987

Scidmore, E.R. *Java, the Garden of the East*, Singapore 1984

Smitinand, Tem *Thai Plant Names*, Bangkok 1980

Steiner, Dr Mona Lisa *Philippine Ornamental Plants*, Manila 1986

Thomas, Arthur *Gardening in Hot Countries*, London 1966

Warren, William and Tettoni, Luca Invernizzi *The Tropical Garden*, London/New York 1991

Whittle, Tyler *The Plant Hunters*, London 1970

Yam, Tim Wing *Orchids of the Singapore Botanical Gardens*, Singapore 1995

Index